Perspectives on Social Sustainability and Interior Architecture

Dianne Smith • Marina Lommerse • Priya Metcalfe
Editors

Perspectives on Social Sustainability and Interior Architecture

Life from the Inside

Springer

Editors
Dianne Smith
Department of Architecture and Interior Architecture
Curtin University
Perth
West Australia
Australia

Marina Lommerse
Marina Lommerse Consultants
Perth
Western Australia
Australia

Priya Metcalfe
Priya Metcalfe Conservation and Heritage Consultant
Perth
Western Australia
Australia

ISBN 978-981-10-1349-2 ISBN 978-981-4585-39-2 (eBook)
DOI 10.1007/978-981-4585-39-2
Springer Singapore Heidelberg New York Dordrecht London

Softcover reprint of the hardcover 2nd edition 2014

Printed on acid-free paper

Springer is part of Springer Science+Business Media (www.springer.com)

Cover photograph by Angela Buckland

The field of Interior Architecture is defined by the way we act, and the opportunities for a shift in, and challenge to, perception.

Foreword

No living thing can be understood without knowing its habitat. This holds true of human society. A sense of space and place is intimately bound up with the human experience of being and becoming—the way in which people find their place as self in society, the way in which they build meaningful relationships with others, the way they individually and collectively construct a 'good life'.

For these reasons it is heartening to encounter such a diverse and challenging collection of essays that connect the field of interior design with the issues of justice in society.

Of course there are competing notions of justice, and competing approaches to human development. Working over many years in the arena of social justice and community development, I have found myself increasingly drawn to Amartya Sen's concept of development as freedom. This is not the false freedom of the market or the fleeting freedom of hedonistic individualism, but rather the agency of individuals and communities to negotiate the world, using their human capabilities to construct and enjoy the 'good life'.

There is no absolute definition or standard of a good life of course. The point is that it can only be found in the mesh of meaning and relationships that we co-create. True human development happens when people create the circumstances for rewarding existence and activity, as they understand it. This is very much a matter of culture and not just driven by the psyche or ego of individuals.

And power relations imbue all of human society and culture. The ideal of justice, however constructed, is constrained by relations of inequality and domination. So I believe there is such a thing as an architecture of justice, both figuratively and literally. In considering the places and spaces where we live out our lives together, political and ethical thinking frames come into the picture.

While these reflections may seem somewhat abstract, the papers in this collection demonstrate that by joining the consideration of place and design to reflection on social relations, we can achieve richer insights that can drive better practice.

In the twenty-first century we seem to have entered into a new world of knowledge discovery, where many of the most exciting insights come not from the authority of a traditional discipline, but from the dialogue that happens at the hubs and intersections of thought—the arenas where different disciplines and approaches, different schools and habits of thinking, come together to collaborate and contend. This collection is a good example of this, and I hope the book will be widely read and its lessons learned and applied.

Tim Costello

Officer of the Order of Australia
Chief Executive, World Vision Australia

Preface

> ... we need a more joined-up way to design for living. This means rethinking everything, including the way humans feed, clothe, shelter, assemble, communicate and live together. It means designing at the level of behaviors, habits, beliefs and language. In short, it entails re-designing design itself. (Wood 2012, n.p.)

One of the main aims of *Perspectives on social sustainability and interior architecture: Life from the inside* is to look at interior architecture through a lens which enables the embedded process, knowledge and skills to be interrogated. Another is to see how interior architecture can add to the debate and discussion concerning worldwide issues of social concern, across disciplines that are directly and indirectly associated with the design professions.

A number of fields in this transdisciplinary spectrum have been emerging over the last 30 years, but in more recent years, the literature has increased. Since the publication of our first edition, *Life from the inside: Perspectives on social sustainability and interior architecture* in 2011, fine distinctions between concepts have been defined. For example, social innovation, social economy, social development, and community development are areas described and discussed in relation to social responsibility and social sustainability. In addition, more common terms in relation to design processes and thinking in interior architecture and related disciplines, including wicked problems, participatory design, co-design, and co-production, are being rediscovered. These terms are applied to discussions of working with people within their contexts, and therefore, engaging with processes that generate insider perspectives of complex and ambiguous situations.

In Chapter 1, An introduction to Social Sustainability and Interior Architecture, we revisit the strong continuing focus on sustainability that pertains to ecological or natural environmental concerns within the literature. Although social, cultural or human aspects are interwoven and open up the discussion, the pendulum still tends to rest with the concepts and rhetoric of the interrelated terms that compose our title—social sustainability and interior architecture.

Therefore the essays contained in this publication continue to inform, extend, and challenge these discussions. We hope that the ideas proposed and questions raised by the authors' contributions will serve to stimulate your practice and research. We also invite information regarding any projects that sit within the realm of social responsibility or social sustainability and interior architecture. In addition, we hope to contribute to the clarification of how interior architecture can play a role in the changing global context.

> …we are driven by two core principles. Firstly, a need for equity of access to good design for all, privileging social justice; and secondly, a need to engage with community through student and staff projects, so that we increase the relevance of interior architecture and foster positive change. (Smith and Lommerse 2010, p. 7)

Dianne Smith
Marina Lommerse
Priya Metcalfe

References

Wood, J. (2012). Why user-centred design is not enough. Core77. http://www.core77.com/blog/articles/why_user-centered_design_is_not_enough_by_john_wood_23465.asp. Accessed 30 Nov 2013.

Smith, D., & Lommerse, M. (Eds.). (2010, April). A permeable field of possibilities. *Permeable boundaries: Interiors—A field of possibilities*. Exhibition catalogue. Perth: John Curtin Gallery, Curtin University.

Acknowledgements

Chapters in this book were double blind reviewed by independent experts in accordance with the Australian Department of Education, Science and Training (DEST) guidelines.

The editors would like to thank and acknowledge the following people for their creative participation, and for igniting debate and development in the compilation, refereeing and editing of *Perspectives on Social Sustainability and Interior Architecture: Life from the Inside*:

- The contributing authors for their enthusiastic response to the intention of the book, and their ideas that shaped the book through the contributed chapters.
- The referees, for considered and constructive assessment of the chapters:
  - Linda Young, Senior Lecturer, School of History Heritage and Society, Deakin University, Geelong
  - Madeline Lester, Past President IFI, International Federation of Interior Architects and Designers, Principal of Madeline Lester + Associates, Sydney, Member of the Order of Australia
  - Salvatore Di Mauro, Senior Lecturer, Queensland College of Art, Griffith University, Brisbane
  - Professor Suzette Worden, Independent Researcher, Suzette Worden Research, Sunbury, Victoria.
- Tim Costello, Officer of the Order of Australia, Chief Executive, World Vision Australia for writing the foreword.
- Christina Houen, *Perspectives'* Production Coordinator, Adjunct Research Associate of School of Media, Culture and Creative Arts, Curtin University, and Editor/Director of www.perfectwordsediting.com, for her dedicated editing and production work, impressive attention to detail and timely advice to authors.
- Sarah Beeck, Lecturer, School of the Built Environment, Curtin University for her research assistance, dedication and contributions to the book preparation.
- Allyson Crimp, Designer, of et-al.com.au, for her graphic work and image editing.

- Jessica Winters, graduate, Curtin University and Tiia Maunu, Interior Designer, for their research assistance and contributions to the book preparation.

We thank Springer for their enthusiastic approach to publishing the book.

Last but not least, we thank Curtin University for support enabling the research and production associated with this book project.

Contents

Contributors

Kelley Beaverford Department of Interior Design, Faculty of Architecture, University of Manitoba, Canada, US

Sarah Beeck Electa Design, Churchlands, WA, Australia

Angela Buckland Durban, South Africa

Marianne Frandsen Designer, honorary member of Danish Designers, Copenhagen, Denmark

Rodney Harber Harber & Associates, Architects, Urban & Regional Planners, Durban, South Africa

Honorary Research Fellow, University of KwaZulu-Natal, Durban, South Africa

Marina Lommerse Marina Lommerse Consultants, South Fremantle, WA, Australia

School of the Built Environment, Curtin University, Perth, Australia

Paul Memmott Aboriginal Environments Research Centre (AERC), School of Architecture and Institute for Social Science Research, University of Queensland, Brisbane, Australia

Priya Metcalfe Priya Metcalfe Conservation and Heritage Consultant, Hamilton Hill, WA, Australia

Priya Metcalfe Conservation and Heritage Consultant, Perth, WA, Australia

Jill Pable Department of Interior Design, Florida State University, Tallahassee, FL, USA

Dianne Smith Department of Architecture and Interior Architecture, Curtin University, Perth, Australia

Penny Sparke Kingston University, London, UK

Lisa Waxman Department of Interior Design, Florida State University, Tallahassee, FL, USA

About the Editors

Dianne Smith is Associate Professor and Head of Program in Interior Architecture at Curtin University. Dianne's research focuses on discriminating design with particular reference to people who are cognitively impaired, the meaning of environments, the impact of colour on experience, and our understanding of place, and design education.

Marina Lommerse is a designer, educator, curator, researcher and strategist with Marina Lommerse Consultants, and Adjunct Professor, Curtin University. She works with creative people and communities to strengthen individual and collective capacity and build supportive environments, mobilizing creative projects that promote positive change in communities, through publications, creative works, and curatorship of exhibitions.

Priya Metcalfe has over 15 years' experience in practise and as a lecturer at Curtin Univerisity, specializing in conservation and heritage. She has prepared heritage assessments and conservation plans for a wide range of commercial and residential buildings. Priya's Master of Philosophy dissertation relates to issues of cultural heritage.

About the Authors

Kelley Beaverford is Associate Professor at the University of Manitoba and Executive Director of Architects Without Borders Canada. She has over 10 years experience as a design educator with a focus on human rights and global citizenship. Kelley has worked on numerous non-profit projects in more than 16 countries.

Sarah Beeck has an extensive career in professional practice and as a Lecturer in Interior Architecture at Curtin University. Her interests include collective memories and objectivity within interiors, and the possibilities of creative production as research.

Angela Buckland photographer, Durban, South Africa, taught for many years in the Fine Arts Department at Durban University of Technology. She currently runs her own private photography school and works as a freelance photographer in Durban. Her photographic work explores deeply personal themes and is held in private and public collections.

Marianne Frandsen has worked on assignments in Denmark and the other Scandinavian countries and the Middle East, including Iraq and Kuwait. She volunteers in organizational work, defining the emerging profession of interior architecture in Denmark as well as internationally.

Rodney Harber is a a practising architect and urban and regional planner, an international development consultant and a Professor of Architecture, Planning & Housing. His award-winning practice, Harber & Associates, has worked mainly in rural developmental projects along the eastern seaboard of South Africa, concerning sustainable settlements, impact of HIV/AIDS, poverty alleviation and community buildings.

Paul Memmott, anthropologist and architect, is Director of the Aboriginal Environments Research Centre (AERC) at the University of Queensland. Current research in the AERC addresses social problems of housing design, crowding, homelessness and family violence in both metropolitan and remote parts of Indigenous Australia.

Jill Pable is Associate Professor in the Department of Interior Design at Florida State University. Her service and research focuses on design for the underprivileged. Through her guidance, two of her students are the holders of a US patent for the design of a cradle for use by families in homeless shelters.

Penny Sparke is Pro Vice-Chancellor (Research and Enterprise) and Professor of Design History at Kingston University, London. She oversees three Faculties and chairs the Research Committee, the Campus Development Board and the Cultural Strategy Group. She has published over a dozen books and numerous articles and book chapters over the last 25 years.

Lisa Waxman is Professor and Chair of the Department of Interior Design at Florida State University. Her current research includes the design of third places and spaces that foster community, as well as housing for special populations. She teaches environment and behaviour, computer-aided design, sustainability, and studio.

List of Figures

List of Tables

Chapter 1
An Introduction to Social Sustainability and Interior Architecture

Dianne Smith, Sarah Beeck, Marina Lommerse and Priya Metcalfe

Abstract Interior architecture is a field that has much to contribute to the discussion and practice of social sustainability. Its potential is not often made explicit, although there are many sources which discuss the design and usage of interiors in relation to material specification, energy consumption and the like; that is, to the relationship to the broader sustainability debate. In contrast, social sustainability captures the human dimension of the sustainability discourse and gives emphasis to people and communities. It is this focus that resonates with interior architecture. We believe that social sustainability is the ability of a society or an individual's lifestyle to continue in a way that suits their needs and those of subsequent generations. The values and spiritual aspirations of the people should be complemented in their interior environment, and the processes and activities involved should respect their history, current needs and future potential beliefs and rituals. We identified that there are three key areas thin social sustainability where the discipline can meaningfully contribute: community engagement, social justice and cultural heritage. Each area is critical to understanding the relationship between social sustainability and interior architecture in the twenty-first century.

Keywords Interior architecture · Social sustainability · Social responsibility · Community engagement · Social justice · Cultural heritage

D. Smith (✉)
Department of Architecture and Interior Architecture, Curtin University, Perth, Australia
e-mail: dianne.smith@curtin.edu.au

S. Beeck
Electa Design, 12 Campus Way, Churchlands, WA, Australia
e-mail: sjbeeck@gmail.com

M. Lommerse
Marina Lommerse Consultants, 11 King William Street,
South Fremantle, WA 6162, Australia
e-mail: marina@marinalommerse.com

Curtin University, Perth, Australia

P. Metcalfe
Priya Metcalfe Conservation and Heritage Consultant, 1 Bradbury Road,
Hamilton Hill, WA 6163, Australia
e-mail: p.metcalfe@bigpond.com

D. Smith et al. (eds.), *Perspectives on Social Sustainability and Interior Architecture*,
DOI 10.1007/978-981-4585-39-2_1,

Introduction

The design and construction of the built environment cannot be removed from the impact that it has on the occupants, the surroundings and the broader community. The act of designing is more than the generation of a physical place. Our discussion concerning the potential role of interior architecture commenced by reflecting on what it is to be an interior architect—the core attributes of our practice and how the relationship between people and their environments is an important aspect of what we do. For example, environments people inhabit and experience have the power to enhance the wellbeing of individuals and their communities. In this unique collection of writing we share new ways of thinking about interior architecture by reconsidering theories, processes and outcomes that relate to how people live now and in the future. The issues raised by the authors are relevant for individuals, government and non-government organizations, professionals, students and communities.

In addition we considered the term 'social sustainability'—a concept that captures how sustainable practices apply to the way people live. However, when applying social sustainability principles to interior architecture, we found the question that needed to be asked was: does social sustainability need to be redefined and developed to suit this discipline? We identified that there are three key areas within social sustainability where the discipline can meaningfully contribute: community engagement, social justice and cultural heritage. Each area is critical to understanding the relationship between social sustainability and interior architecture in the twenty-first century.

Social sustainability captures the human dimension of the sustainability discourse, and to define it in terms of interior architecture it was necessary to think about the terms 'sustainable development' and 'sustainable communities' as well. Sustainable development is: '… development sustainable to ensure that it meets the needs of the present without compromising the ability of future generations to meet their own needs' (World Commission on Environment and Development 1987, p. 240), whereas sustainable communities are characterized by a strong connection to place, building capacity and resilience to external forces. The former is an eco-centric approach, emphasizing the effective use of resources, and the latter is an anthropocentric approach, focusing on human relationships, needs and quality of life (Kearns and Turok, as cited in Jarvis et al. 2010).

It is the emphasis on people and communities that resonates with interior architecture. As a result, an anthropocentric focus is adopted in this book or, as Helen Jarvis puts it, the connection between people's 'livelihood and environments' (Jarvis et al. 2010, p. 4). We believe that social sustainability is the ability of a society or an individual's lifestyle to continue in a way that suits their needs and those of subsequent generations. The values and spiritual aspirations of the people should be complemented in their interior environment, and the processes and activities involved respect their history, current needs and future potential beliefs and rituals. Access to such environments should not be at the expense of others. Social sustainability recognizes that the environment and its design facilitate relationships and experiences fundamental to who we are as people and as a society; and that the basic human rights to dignity, security and shelter are met now and into the future.

Recently, other authors have raised questions concerning the definition and practice of social sustainability. Constanza Parra stresses that social sustainability must go beyond 'equity and justice' to include the integration of knowledge about how people live, build societies and address challenges together. In addition he raises the need for the integration of development and innovation with the social aspects, as both these concepts are concerned with governance. The intention is to foster 'an innovative logic of interactivity seeking a transformation in social relations oriented to define and satisfy human needs', as well as to recognize and enable a 'nature-society relationship' (Parra 2013, p. 142).

Such observations, however, raise additional questions regarding social sustainability. Should social sustainability always be about needs, about development, and/or concern innovation? Literature reviews such as Vallance, Zerkins, and Dixon's seek to define social sustainability as a 'particular branch of sustainable development' (Vallance et al. 2011, p. 342). In addition, they categorize the social sustainability literature as development, bridge and maintenance sustainability. Vallance et al. state that development sustainability appeals as it marries people's needs with bio-environmental management and economic development and growth. However, they also identify that later studies do not indicate that the poor are necessarily assisted significantly; and raise the point that often, sustaining 'their present societal arrangements' is an 'unsustainable thought' (Vallance et al. 2011, p. 343). The focus on third world over first world nations is also raised—and they ask if basic needs can and should be met across all countries. In contrast, bridge social sustainability 'attempts to harness human potential so as to generate improved environmental outcomes' (Vallance et al. 2011, p. 344). This may be in ways that seek to transform concepts of people-environment relationships (lifestyle) or through traditional approaches that deal with environmental information while potentially fostering change (by technology for example). Both categories are still couched in biological environmental language although they define the social.

The third category arising from their review, maintenance social sustainability, draws on more recent literature concerned with 'social cultural preferences and characteristics, and the environment, [that] are maintained over time' (Vallance et al. 2011, p. 345). Importantly, the possible disruptive impact of 'eco-friendly' interventions on a community is highlighted; and the authors state that a sustainable city is one where people 'want to live' (Vallance et al. 2011, p. 345). The tension between bio-physical interests or interventions and what people will tolerate is also discussed, and they propose the need to examine the fine grain 'habits as well as the more conscious preferences' of people or community involved (Vallance et al. 2011, p. 346).

Socially responsive design is posed as an alternative to socially responsible design per se by Thorpe and Gamman (2011). Detail is given to approaches that can undermine the principles of co-design: for example, maternalism, fraternalism, paternalism. They stress that the term responsibility may shift the perception that the designer is an expert controller, whereas they consider that the designer is a facilitator. Thorpe and Gamman (2011) cite Hillgren et al.'s (2011) listing of a number of perceived limitations that design has in the field of social innovation:

- lack of economical and organisational skills;
- inabilities in driving the implementation process;
- high cost of design consultants;
- often do not have a long-term commitment in projects;
- superficiality of some proposals which ignore evidence, and field and historical experiences, and which have a tendency to "re-invent the wheel". (Hillgren 2011, as cited in Thorpe and Gamman 2011, p. 224)

And as such, designers cannot (and should not claim that they can) be responsible for socially focused projects. Instead they call for an alternative approach, co-design. The skills and knowledge of interior architects, we propose, can play a purposeful and significant role in alternative approaches. The chapters within this book aim to communicate such processes and beliefs, which are centred around projects pertaining to social responsibility.

Parra (2013) uses the term social to convey how people 'converse and interrelate across changing patial-temporal contexts … the institutional, political, and socio-cultural forms through which human life is experienced and (re)produced in different places of this world' (as cited in Parra 2013, p. 146). Even so, the discussion still focuses on biological or ecological matters—although states, nature and cultural matters cannot be separated. To provide some balance to the discussion, through this collection of essays, we would like to stress the person-focused view as a counterpoint to the well documented biological/resource focus in the sustainability debate within design disciplines. For example, in regard to interiors, much of the sustainability discussion deals with reuse and recycling of materials, passive design, energy load, embodied energy and life cycle costing (for example, Usal 2012; Máté 2007; Thibaudeau 2008; Hankinson and Breytenbach 2012). Sorrento points out that an ecological approach has replaced the original (and more mechanistic) approaches by looking at the interdependency of people and nature. As a result the projects are 'based on strategies of adaptation, resilience, and regeneration that build, rather than diminish social and natural capital' (du Plessis 2012, as cited in Sorrento 2012, p. ix). In addition, Sorrento (2012) calls for the integration of evidence based design to help define more purposefully the criteria for sustainability and qualitative dimensions of human experience. The consideration of changing client beliefs and values about sustainability may form part of the human dimension.

Again questions are raised. Why is the focus so biased toward the eco-centric approach to sustainability? Is the experience of humans in this field limited to human behaviours around consumption and/or designing? Are evidence based approaches, which strive to measure and quantify, the best approaches to situations that are complex and involve human beliefs, values and lifestyles? Or can we benefit from pluralist approaches and does this shift imply a need to develop more inclusive and exploratory approaches for social sustainable design?

> In considering the role of co-design in socially responsible and sustainable design initiatives, Melles et al. note, among a number of objectives for design (and designers), the need to avoid the passivity of consumers. Not surprisingly, at a time when there is pressure for less consumption and greater preservation and distribution of resources, discussion about the roles and responsibilities of design is clearly moving far beyond the 'socially useful design' arguments raised by Papanek (1971/1984) 40 years ago, and by Whiteley (1993) almost 20 years ago. (Thorpe and Gamman 2011, p. 218)

Traditionally, interior designers have referred to inhabitants of spaces as users. Sanders and Stappers (2008) note that the American concept of a user-centred design approach reflects the person as a subject, and that more recently, the occupants or people are conceived as partners. Partner processes are European led (Sanders and Stappers 2008, p. 5). As partners, a person or community is part of the design process, and is able to provide essential aspects of a project, including ideas and expertise; that is, co-design is not just between differing designers, for all people are able to engage. The impact of the way we design and our associated processes and collaborations also raise questions, because: '[b]ringing co-creation into design practice will cause a number of changes to occur. It will change how we design, what we design, and who designs' (Sanders and Stappers 2008, p. 16).

Social Sustainability and Interior Architecture

Perspectives on social sustainability and interior architecture: Life from the inside positions the debate in relation to our practices through three key areas. We define each theme in relation to social sustainability as follows.

Community engagement is predicated upon the consent and active participation of the community whose social capital, processes or environments are being considered, thereby playing a vital role in social sustainability. Here interior architecture works with the community. Through community engagement we see the potential to identify the needs—and more importantly the assets or social capital—of people and communities and so we can begin to develop a range of options to consider. These options may be environments, ways of working, and transferable and discipline-specific skills and knowledge that are able to reflect that community, rather than a homogeneous approach to all products or processes. Community engagement allows for the exchange and sharing of social capital between people in the community and interior architecture practitioners, students, educators and researchers (Bell and Wakeford 2008).

Within the broad field of community engagement are numerous relationships and projects that focus on issues of equity of access to basic human rights that pertain to social justice. To have the capacity to access these rights is an aspiration for all, but the ability to do so is often taken for granted. In order for all people to have an authentic existence, their ability to have access to resources and practices to facilitate their needs and requirements, while maintaining their dignity and identity in an equitable way, is fundamental.

Social sustainability captures the relationship between a culture and how it operates as a society in its immediacy, while embracing its longevity and evolution; social justice issues are embedded in this relationship. Interior architecture deals with both processes and the environments that are implicated, and if harnessed can foster positive change at a micro or personal level. Veronique Vienne (2003, p. 245) states that 'we are about to redefine the very nature of design' and there is a pressing need to consider how interior architecture relates to the quality of life for those who are without, disadvantaged or non-mainstream.

While basic human needs of clean water, shelter and sanitation are crucial to existence the world over, people's lived experiences embrace other aspects of life. Aspects such as community, cultural, social and ritual behaviours are equally important in the minutiae of daily life. The sites of these aspects are both natural and man-made, and encompass work, social interaction, traditional religious or cultural rituals, and domesticity. It is the constructed environments that we are concerned with here, and their social sustainability through continued use—sometimes with evolution over time that reflects shifting social mores and changing technologies—and at other times through adaptive re-use of what would otherwise become redundant sites, settings, buildings and their interiors.

Understanding the importance, or otherwise, of these places comes through recognition of the value these places may have to the local or wider community, in their associations to past generations and the ongoing desire to retain the memories embodied in buildings and their interiors.

An integral part of the understanding of built cultural heritage is through a rigorous approach to a place's history, its current use and condition, and the identification and retention of particular values. A fundamental part of this process is identifying the values of places through community consultation with stakeholders—be they government, business or occupants and users—to understand to whom a place may or may not be important and why. Designers of the built environment are and should be concerned with this process before embarking on conservation, restoration, adaptive re-use and/or interpretation.

As an extension of that knowledge, cultural heritage may not always concern itself with existing early buildings. It also includes working with communities to integrate an understanding of the essence of their cultural heritage within contemporary environments, to enable a sense of continuity, relevance and belonging.

Unfolding the Book

The chapters that follow present possibilities for interior architecture to be involved in social sustainability in a meaningful way. They celebrate important work underway, and question why the discipline is not leading in areas where our core competencies can obviously make a difference. The book outlines the possibilities for action and involvement into the future.

The terms interior architecture and interior design, and interior architect and interior designer are used in various parts of the world to describe the discipline area and its practitioners. The terms are used interchangeably in this book.

The book contains perspectives from an anthropologist, as well as interior practitioners, researchers and educators. Their case studies provide a varied, international overview—set in various urban, rural and remote places in the United Kingdom, Indonesia, Africa, Europe, Canada, the United States and Australia—in their engagement with local contexts that are set within larger socio-economic-political contexts and issues.

Community Engagement and Interior Architecture

Community engagement is conceptualized as 'ways of working', and in this section, we explore frameworks that make it possible for people to be involved in shaping and managing their environment. These 'ways of working' can be, and sometimes—but not always—are used in the development of strategies, processes, design and construction of the environments debated in the second and third sections: social justice and interior architecture, and cultural heritage and interior architecture.

The second chapter, Working together: interior architecture creating with the community, begins with three premises: that community engagement is a place to learn and exchange knowledge; that it embodies ways of working with a community to identify priorities, dream of possibilities, make shared decisions, and make things happen; and that the interior architecture community can enrich and open up opportunities for other communities. The chapter goes on to define community and conceptualize community engagement in relation to design, and the socio-political forces that affect community. Next, examples from fields with established and emerging structures for community engagement are critiqued from two perspectives: the ways they engage and their success in progressing the interests and social sustainability of the communities concerned. The chapter concludes with proposed strategies for interior architecture to create with the community.

Featured in the third chapter, More than bricks and mortar: lessons in design and world citizenship, is the illustration of a framework, possibilities and challenges of a service-learning project utilizing community engagement principles. Located in Ghana and involving Canadian students and faculty, the design–build project was a partnership between the community, the students and the university. Described are themes of engagement, reflections, challenges and limitations. The chapter concludes with the possibilities and benefits for such community engagement projects to facilitate cross-cultural experiences.

Social Justice and Interior Architecture

This section explores how interior architecture can participate in the debate and in programs that address social equity and social justice for all members of society in relation to social responsibility or sustainability. The section challenges interior architects to use their creative talent and processes to envisage new ways of tackling old problems, that will result in strategies to develop environments in which to live, work and play, where the quality of our and others' lives are enhanced, not lessened.

The fourth chapter, Interiors can address Social Justice: fact or fiction? begins by defining social justice and how interior architecture can engage with the concept. By examining professional organizations and their mandates, exploring examples in practice, and surveying university programs engaging in social justice, the kinds of involvement the discipline has or could have with social justice are illuminated. The chapter focuses on activities that explicitly involve core interior architecture competencies, and/or engage interior architects in order to ascertain how active the discipline is in the area. The chapter concludes with an outline of the role interior architecture can play in social justice in the future. A critique of the contextual issues that facilitate and/or inhibit interior architecture practitioners—addressing social justice as an area of practice—highlights areas for future research.

The following Chapters 5–8, provide four perspectives from key players in practice, to emphasize the potential interior architecture has to engage with social justice. The first perspective, presented in Chapter 5, The importance of interior design professionals' engagement in policy development to effect change for underprivileged individuals, is from the United States, and addresses interior architects' involvement in policy development for underprivileged individuals. A shift in thinking about standards for low-income housing, allowing for the needs and identities of residents, opens an opportunity for the discipline's involvement in policy. The sixth chapter, Interior design by expediency, illuminates a case study of marginalized migratory workers' hostels in South Africa, describing the undignified conditions the workers encounter and their battle to express identity and create their own space. It looks at how interior architecture could provide for these real needs of men and women in a disadvantaged situation. Chapter 7, Inside the remote-area Aboriginal house, discusses how culturally distinct practices should inform the design of housing for Aboriginal Australians in order to mitigate indirect discrimination, which often occurs in the homogenized housing models that have been a feature of government policies. It identifies that current housing for remote Aboriginal communities does not cater to socio-cultural needs, and discusses how, by developing a sensitive cross-cultural methodology, interior architects have the potential to make a meaningful contribution. The fourth perspective, presented in Chapter 8, Too 'nice' to be socially responsible: reflections of a Danish interior practitioner, discusses a Danish guerrilla design movement termed 'social indignation'. The projects presented use provocative design of social spaces in order to challenge inhumane and rational treatment of spaces. Interior architects are challenged to take an active role in confronting current policy through the innovative approaches currently being created by artists.

Cultural Heritage and Interior Architecture

In this volume, social sustainability also refers to cultural heritage. Working with cultural heritage is concerned with understanding the importance of particular places to people, in aspects of their daily lives such as community, culture, and social and ritual behaviours. Interior architecture can create and/or adapt places through recognition of the values these places might have to the local or wider community, in their associations to past generations and the ongoing desire to retain the memories embodied in buildings and their interiors. The chapters that feature in this section consider the understanding of built cultural heritage through the history of places, their current uses and condition, and the identification and retention of particular values.

Chapter 9 focuses on the issues around socially sustainable, adaptive re-use through a critique of current practice working with places that have been identified as having cultural heritage significance. There are a substantial number of places in Perth, Western Australia that have been, or should be recognized as having value to the community. While the exterior of these early buildings typically undergo some modification over time, generally the interiors of these places are subject to more frequent modification and adaptation through changes in use, or to improve amenity. The chapter discusses an industrial site in the process of adaptation for re-use as an exemplar of socially sustainable cultural heritage.

The tenth chapter focuses on the Modern interior and the question of domesticity in relation to the restoration, conservation or adaptive re-use of buildings. The richness and meanings of the interior past and present are revealed technologically, visually, materially, spatially, psychologically and socio-culturally. The shifting socio-cultural influences and how they impact the interior—relative to domesticity—are explored as to how interpretation relevant to the present uses may be expressed. The chapter concludes with how the field is evolving and the important duty interior architects and allied professions have to search for new interpretations and approaches, in order to promote social sustainability.

Provocation

Numerous issues arising in the transdisicplinary field of social sustainability and interior design are reported in the three parts of *Perspectives on social sustainability and interior architecture: Life from the inside*, forming a collective dialogue. Many of these issues may have been addressed—at least in part—by others who are not included in this volume. As part of the continuing discourse and reflection on interior architecture, we are interested in making contact with such people, and disseminating information on this field as it evolves. We are interested in both retrospective and future projects for the 10 years prior to, and 10 years onward from, the book publication date.

Debate in this area of interior architecture is both important and timely. Our research has revealed a cross-section of projects and players, but key questions still remain, and processes still need to be developed. Our provocation for debate is that interior architecture is a field through which practitioners, educators and researchers can make significant contributions to social sustainability, social justice, community engagement and cultural heritage, from their core skills and knowledge base.

Questions for debate and research are: How can the core skills of interior architecture be integrated into mainstream practice to ensure social justice through design? How do design and construction practices involving client consultation need to evolve to reflect designing *with* the community? How can cultural heritage awareness act as a driver for contemporary design innovation? Is social sustainability an imperative within ethical interior architecture practice for the twenty-first century?

In our quest for answers we invite all—theorists, practitioners, and the general community—to contemplate, to discuss and to act.

References

Bell, B., & Wakeford, K. (Eds.). (2008). *Expanding architecture: Design as activism*. New York: Metropolis Books.

Hankinson, M., & Breytenbach, A. (2012). Barriers that impact on the implementation of sustainable design. http://cumulushelsinki2012.org/cumulushelsinki2012.org/wp-content/uploads/2012/05/Barriers-that-impact-on-the-implementation-of-sustainable-design.pdf. Accessed Nov 2013.

Hillgren PA, Seravalli A, Emilson A.(2011). Prototyping and infrastructuring in design for social innovation. *CoDesign*, *7* (3–4), 173–187.

Jarvis, H., Pratt, A., & Cheng-Chong Wu, P. (2010). Understanding social sustainability: Key concepts and developments in theory and practice. In T. Manzi, K. Lucas, T. Lloyd Jones, & J. Allen (Eds.), *Social sustainability in urban areas: Communities, connectivity and the urban fabric* (pp. 1–28). London: Earthscan.

Máté, K. (2007). Using materials for sustainability in interior architecture and design. *Journal of Green Building, 4*, 23–38. doi:http://dx.doi.org/10.3992/jgb.2.4.23.

Parra, C. (2013). Social Sustainability: A competing concept to social innovation? In F. Moulaert, D. MacCallum, D. Mehmood & A. Hamdouch (Eds.), *The international handbook on social innovation: Collective action, social learning and transdisciplinary research* (pp. 142–154). Cheltenham: Edward Elgar Publishing.

Sanders, E. B. N., & Stappers, P. (2008). Co-creation and the new language of design. *CoDesign*, *4*(1). http://www.maketools.com/articles-papers/CoCreation_Sanders_Stappers_08_preprint.pdf. Accessed Nov 2013.

Sorrento, L. (2012). A natural balance: Interior design, humans, and sustainability. *Journal of Interior Design*, *37*(2), ix–xxiv.

Thibaudeau, P. (2008). Integrated design is green. *Journal of Green Building, 3*(4), 78–94. doi:http://dx.doi.org/10.3992/jgb.3.4.78.

Thorpe, A., & Gamman, L. (2011). Design with society: Why socially responsive design is good enough. *CoDesign: International Journal of CoCreation in Design and the Arts, 7*(3-4), 217–230. doi:10.1080/15710882.2011.630477.

Usal, S. (2012). Evaluation of product consumption understandings of interior architecture students in terms of sustainability. *Procedia-Social and Behavioral Sciences, 47,* 351–356. doi:10.1016/j.sbspro.2012.06.662.
Vallance, S., Perkins, H. C., & Dixon, J. E. (2011). What is social sustainability? A clarification of concepts. Geoforum, *42*(3), 342–348. doi:38.3016/f.geoforum.2011.01.002.
Vienne, V. (2003). In a continuous state of becoming: Design responsibility tomorrow. In S. Heller & V. Vienne (Eds.), *Citizen designer: Perspectives on design responsibility* (p. 245). New York: Allworth Press.
World Commission on Environment and Development. (1987). *The Bruntland report*: *Our common future*. Oxford: Oxford University Press. http://www.worldinbalance.net/pdf/1987-brundtland.php. Accessed 10 March 2011.

Part I
Community Engagement and Interior Architecture

Interior architects realize that issues of social sustainability cannot be solved through physical spaces alone. Nor can a transformation be facilitated through a 'one-size-fits-all' approach. Community engagement, we envisage, is simultaneously a philosophy, a way of working, an attitude and action, whereby it is possible for people to be involved in shaping and managing their environment. In this section, the intersection between Community Engagement and interior architecture is investigated. The role of citizen designers is in part to build the capacity of individuals and the group, as well as to facilitate strategies to improve the built environment. This positions interior architects as problem identifiers as well as problem solvers (Peterson 2008).

Community engagement is a vehicle for citizens to communicate about their worlds and share their visions. This requires cultivating a sense of a shared citizenship, where people identify priorities, dream of possibilities, make shared decisions, and make things happen. The diversity of participants provides a key resource—transcultural, interdisciplinary and intergenerational.

Participants function as both teachers and learners. Citizen designers are both members and enablers of community; they can play multiple roles, including those of social thinkers, educators and human rights observers (Aeschbacher and Rios 2008). Mutual respect and sharing of skills and knowledge from inside and outside the community can bring a richness and synergy unparalleled in other practice modes.

The following slogan, from a poster for The People's Hurricane Relief Fund in New Orleans, captures the idea of Community engagement; that is, the people with the least amount of power must be central to the decision-making process that impacts their everyday personal and political lives, but especially their educational lives.

'Nothing about Us, without Us, is for Us.' (The Electronic Intifada n.d.)

References

Aeschbacher, P., & Rios, M. (2008). Claiming public space: The case for proactive, democratic design. In B. Bell & K. Wakeford (Eds.), *Expanding architecture: Design as activism* (pp. 84–89). New York: Metropolis Books.

Peterson, J. (2008). Mobilising mainstream professionals to work for the public good. In B. Bell & K. Wakeford (Eds.), *Expanding architecture: Design as activism* (pp. 94–103). New York: Metropolis Books.

The Electronic Intifada. (n.d). http://electronicintifada.net/v2/article6554.shtml. Accessed 15 Feb 2011.

Chapter 2
Working Together: Interior Architecture Creating with the Community

Marina Lommerse

Abstract Community engagement partnerships and projects have gained momentum in built environment and design fields, and the potential is great for interior architecture to engage with the community in the context of social sustainability. However, there is little knowledge in the field of how to go about it. Hence there is a need to explore suitable frameworks and methods. Using case studies and reflection on 30 years of practice, I argue that the interior architecture community, using core knowledge, can enrich and open up opportunities for other communities, and it is imperative to let communities, stakeholders, and other fields know what interior architecture can contribute to the public good. This chapter offers insights on how we may integrate community engagement into interior architecture education, internship and interior practice. And how we can take action to develop structures, methods and case studies, so that knowledge-action opportunities are built.

Keywords Interior architecture · Architecture · Planning · Community development · Social sustainability · Community engagement · Participative design · Design activism · Co-design

---

M. Lommerse (✉)
Marina Lommerse Consultants, Fremantle, Australia
e-mail: marina@marinalommerse.com

School of the Built Environment, Curtin University, Perth, Australia

D. Smith et al. (eds.), *Perspectives on Social Sustainability and Interior Architecture*,
DOI 10.1007/978-981-4585-39-2_2, © Springer Science+Business Media Singapore 2014

Fig. 2.1 'A community is a group of people who have something in common' (Bullock and Trombley 2000, p. 145). Tanganyika circa 1957. (Photographer Petrus G. W. A. Lommerse)

Introduction

In this chapter I explore the potential for interior architecture to engage with the community in the context of social sustainability. My objectives are: to engage the profession in taking a leadership role in the arena; to uncover ways in which interior architecture might contribute; and to let communities, stakeholders, partners, and other fields know what interior architecture can contribute.

Community engagement, and its relationship with interior architecture, has informed my most successful endeavours. The principles which evolved through my work are: firstly, to respect each person involved for what each of them can contribute, regardless of experience or education; and secondly, to create synergy between the parties, which leads to transformation (Fig. 2.1).

I define community engagement as a cooperative process of working with people to address their wellbeing, crossing disciplinary boundaries, and using multiple knowledge from inside and outside the community. Community engagement is predicated on the consent and active participation of the community whose social capital, processes or environments are being considered, thereby playing a vital role in social sustainability (Hendler-Ross and Hendler-Ross 2008). Bryan Bell argues that 'the process of creating the built environment can allow communities and individuals to improve and celebrate their lives. It can help solve their struggles by reshaping their existence' (Bell 2008, p. 15).

Community engagement has been part of my practice, research and teaching for two decades, and I have noticed a growth in its application across many fields during that time. In a sense, we have seen a rediscovery of community, and the significance of its values and power. My purpose in this chapter is to facilitate this development of practice and discourse in the design field by presenting some ideas, frameworks and principles for community engagement. These are drawn from a review of published and unpublished case studies from those whose work is engaged with community; in addition, I reflect on how community engagement has evolved in my life and career as an interior architect and educator. This reflection suggests possible answers to questions such as: what is the value of community engagement in relation to social sustainability? How can interior architecture be meaningfully involved?

Community Engagement and Social Sustainability

An understanding of Community Engagement, its relationship to sustainability and the forces which have made it a popular notion, is required to contextualize this study. Sustainability is concerned with balancing social, economic and environmental needs, which vary from region to region, community to community and person to person. Social sustainability, as defined by the editors of this collection (see Chapter 1) refers to the human dimensions of the sustainability discourse—relationships, needs and quality of life within regions and communities. Therefore, by identifying the concerns of a community, and the contextual and cultural practices that it sustains, we can start to define what is 'sustainable' for that particular community.

Evolution of Community as a Central Player

Interior architecture does not operate in a vacuum. The communities we work with, and their social sustainability, are affected by policy, politics and locale. Therefore, an understanding of contemporary concerns, how these relate to community, and the effect of politics and policy is required in relation to global, national and local contexts.

The concept of community has evolved, particularly since the nineteenth century. A perceived loss of community was experienced during the significant transition from a rural agrarian society to an urban industrialized one, resulting in the shift from extended families living together, and longevity of residence. Overall, the advent of the information age, a global economy and increased mobility have resulted in the breaking-up of nuclear and extended families and communities. The idealized forms of pre-industrial relationships were characterized by 'a high degree of personal intimacy, emotional depth, moral commitment, social cohesions and continuity in time' (Scott and Marshall 2009). In contrast, in the first half of the twentieth century, cultural dispossession was rife, with dominant cultures seeking to subsume the colonized ones. But within decades change was imminent: '… the 1960s movement for social equality for minorities and indigenous people started a renaissance', implying an emerging sense of responsibility to community (Lommerse 2009, p. 3).

In the 1960s, 'participative design' became common in planning and architectural schools and practices. This generated many projects and spawned research and methods in 'community design' that form the basis for today's practices. However, '[b]y the 1980s … most of the community design centers were purged from academia, and the style wars … absorbed all that student energy' (Badanes 2008, p. 249).

Community as a metaphor emerged in the 1980s, used by political philosophers who became known as Communitarians. One insight of Communitarians is the recognition that all communities are different and are based on a unique set of relationships and experiences. This point is manifested in the debates on indigenous affairs, multiculturalism, and reconciliation in Australian politics. The term community implies that we need to support a wide range of communal organizations, which could include, for example, same-sex couples, religious groups, charitable organizations, Indigenous and non-Indigenous groups. The politics of community is complex and we must recognize that these communities may often come into conflict with one another. A rise in Communitarianism in the 1990s, associated with Amitai Etzioni, a sociologist, called for the rebuilding of community to restore the health of American society (Little 2007). Bryan Bell, Cynthia E. Smith and Tony Fry, experts in the field of design and sustainability, discuss how designers can work with communities in order to create positive change (Bell 2008, p. 15; Smith 2007; Fry 2009). Case studies in the next sections flesh these ideas out, and point to ways that interior architecture can contribute to this change.

In regard to community there are different agendas in various societies. For example, Adrian Little, an expert in politics and social theory, illuminates Australia's contemporary political position:

> In Australia 'community' is a central notion in political debate … it is not unusual to see community being promoted to represent the whole of Australia: 'community renewal' is concerned with rebuilding the nation from below. This kind of model sometimes implies that certain values and principles underpin the Australian way of life … and the idea of community embodies and reinforces them … [A]lternative community discourses recognize the diversity of Australian society and the need to understand the multiplicity of different groups. This model of community feeds into debates on multiculturalism and, in particular, the political rights and demands of Indigenous communities. (Little 2007, p. 121)

Concepts of community have been debated by sociologists for almost a century, culminating in a reawakening to the idea in the late twentieth century, spurred by many factors, including continued social activism, expression of cultural diversity, the increase in refugees and homelessness, as well as recognition of the need for sustainability and concerns about the perceived loss of community as a cause of social disintegration. In planning and architectural schools, there is a renewed interest and involvement in community engagement and community design. The pendulum has swung back to social justice issues in architectural education; inspired by groups like Rural Studio, students are demanding that their education have meaning and a hands-on 'service component', and many other creative areas are also picking up the gauntlet to aid in issues of community (Badanes 2008).

Over the last 20 years, community has become one of the pivotal platforms of contemporary politics in westernized countries (Little 2007). A popular concept is 'community building', that is, practices meant to increase social justice and wellbeing and reduce the negative impacts of otherwise disconnected individuals. Community building, sometimes referred to as 'community renewal', looks to build both the social networks and the built environments people engage with, resulting in gentrification and renewal of neighbourhoods, which can lead to continuity, but can also destroy or manipulate people's lifestyles and rituals (Community Renewal International n.d.).

Therefore, in reviewing how interior architecture can engage with community, two aspects need to be considered: effective ways to engage, and the benefits of engaging, associated with the purposes and outcomes defined by the community concerned.

What is Community Engagement?

Community engagement is a broadly used term, yet in my experience, within interior architecture circles, it is narrowly understood. The concept and its application can be much more powerful in interior architecture than currently conceived. For example, those who, in the course of my research, responded to my call for case studies for community engagement and interior architecture, exclusively identified projects with the following limited profile:

- design only or design–build projects
- projects for the socially disadvantaged
- community understood in terms of 'community of location'
- pro bono, voluntary or service learning projects.

I believe we need to redefine and expand the conceptualization of community engagement and interior architecture in order to open up career paths and practice areas and to influence policy development for communities.

Community is spoken about in terms of values, such as solidarity, commitment, mutuality and trust, as well as 'communities of location' and 'communities of interest' (Cohen 1985). "'[C]ommunity' involves two related suggestions: that the members of a group have something in common with each other; and the thing held in common distinguishes them in a significant way from the members of other possible groups. Community, thus, implies both similarity and difference" (Cohen as cited in Smith 2001, n.p.).

Community organizations range from informal family or kinship networks, to more formal incorporated associations, political decision-making structures, economic enterprises, or professional associations on a local, national or international scale. Broad principles underpin engagement with them, such as:

- consensus on agenda, procedures and effectiveness
- representativeness and inclusiveness
- deliberation
- capability and social learning

Fig. 2.2 First Nations Elder speaking in front of the 'talking wall', which provided an inclusive and visible listening device. (Photograph by McFarland and Marceau Architects, used with permission)

- decision responsiveness
- a need for clarity of objectives, and for legal, linked and seamless processes
- transparency and enhancement of trust. (Petts and Leach 2001, as cited in The Department of Sustainability and Environment 2005, p. 12).

Juanita Brown and David Isaacs have developed the 'Six Cs' model; these are basic principles to guide Community Engagement. These may be seen as filters to measure the quality of the functioning of the community, as follows:

1. Capability: the members are capable of dialogue.
2. Commitment: mutual benefit beyond self interest.
3. Contribution: members volunteer and there is an environment that encourages members to 'have a go' or take responsibility or risks.
4. Continuity: members share or rotate roles and, as members move on, there is a transition process that sustains and maintains the community corporate memory.
5. Collaboration: reliable interdependence. A clear vision with members operating in an environment of sharing and trust.
6. Conscience: embody or invoke guiding principles/ethics of service, trust and respect that are expressed in the actions of the community. (Brown and Isaacs 1994, as cited in The Department of Sustainability and Environment 2005, p. 12).

To illustrate, for the First Nations House of Learning (University of British Columbia, Canada) a method was needed to gather appropriate material to inform all aspects of the project, and provide a way to debate and inform consensual decision-making. Over years of practice, McFarland and Marceau Architects has evolved an inclusive and visible process to collaborate and have a dialogue with the First Nations community—the 'talking wall' (Fig. 2.2).

> A range of cross-tribal, multi-generational people, oral traditions, alternate ways of working, different languages and essential stories needed to be involved. Through a series of open meetings and the sharing of food and ceremony, a framework for gathering information was established. Each meeting focused on a topic: identity, function, site and image … Participants explored questions like: How do we want to be seen on the campus? What is the right site for us?… Ideas related to topics were put on a card on the wall for all to see. This 'talking wall' stimulated discussion. At subsequent meetings, ideas on the new topic would build onto the existing wall. The talking wall worked because it was visual and immediate. It drew out people of diverse backgrounds and ages. (McFarland 2004, as cited in Lommerse 2009, p. 10)

Additional principles that partners in community engagement need to consider include:

- a commitment to reciprocity that includes stating what you require of the community and delivery of what you will provide in exchange
- establish what you are promising as part of the engagement process, this could include provision of information or feedback on how contributions have influenced decisions, through to implementation of stakeholder decisions
- genuineness in building relationships with community and other stakeholders
- valuing the opportunities that diversity has to offer. (Petts and Leach 2001, as cited in The Department of Sustainability and Environment 2005, p. 14)

Contemporary concepts of community building have led to the development of theories and practices for engaging communities that take on various forms in different countries. In Australia, the state governments of Queensland and Victoria, for example, have established strategies for Community Engagement and community building. The Victorian Department of Sustainability and Environment (DSE) and other like organizations have built a body of knowledge in the field, and their publications, toolkits and online resources can aid communities and those working with them. The DSE defines the principles as follows: '… engagement' is used as a generic, inclusive term to describe the broad range of interactions between people' (DSE 2005, p. 10). Depending on the situation, 'engagement' can cover consultation, communication, education, participation or working in partnership. 'It can include a variety of approaches, such as one-way communication or information delivery, consultation, involvement and collaboration in decision-making, and empowered action in informal groups or formal partnerships' (DSE 2005, p. 10).

> The word 'community'… is … a term used to define groups of people; whether they are stakeholders, interest groups, citizen groups, etc. A community may be a geographic location (community of place), a community of similar interest (community of practice), or a community of affiliation or identity (such as industry or sporting club).
> 'Community engagement' is therefore a planned process with the specific purpose of working with identified groups of people, whether they are connected by geographic location, special interest or affiliation, to address issues affecting their wellbeing. (Queensland Department of Emergency Services 2001, as cited in DSE 2005, p. 10)

To illustrate, examples of 'community of location' and 'community of interest' follow. Geographic location is most commonly thought of when speaking of community, especially in built environment disciplines. Communities of location refer to a group of people living in the same locality and under the same government. It can also refer to the district or locality in which such a group lives. Often referred to as geographic communities or communities of place, they range from the local neighbourhood, suburb, village, town or city, region or nation to the planet as a whole. Community can also refer to nations or a group of nations with a common history, or common economic or political interests (Henri and Pudelko 2003).

In contrast, a 'community of interest' refers to a group of people having common interests, for example, a professional community with the same or related occupations. Some of those members may join a professional society, making a more defined and formalized group. These are sometimes known as communities of practice. A distinct segment of society, for example the gay community, can be viewed as communities of culture or identity and can range from the local clique, subculture, ethnic group, religious, multicultural or pluralistic civilization, to the global community cultures of today. They may be included as communities of need, such as people with disabilities or homeless people (Tropman et al. 2006).

The DSE strategy document goes on to say:

> Linking the term 'community' to 'engagement' serves to broaden the scope, shifting the focus from the individual to the collective, with associated implications for inclusiveness, to ensure consideration is given to the diversity that exists within any community. Community engagement can take many forms and covers a broad range of activities. (DSE 2005, p. 10)

People are striving for empowerment; this means they need to be part of the discussion, the decisions, and the building of networks and environments. Community engagement offers the vehicle through which this can be achieved. If people are to feel proud of their space they need a sense of ownership and buy-in, which engagement brings.

> Key cultural theorists, hooks (cited in Stevens and Acland 1999), Hampton (1995) and Krinsky (1996) have identified important aspects in people-environment design: people need to define their own space; the lack of involvement is dehumanizing; and involvement provides a positive self-image and a sense of power. (Lommerse 2009, p. 7)

Proud and engaged people lead to thriving and sustainable communities.

Reimagining Interior Architecture and Community Engagement

The discipline of interior architecture can be described as a community of practice. Its members include interior architecture practitioners, students, educators and researchers. The professional organizations that link the interior architecture community at local, national and international levels are overviewed in Chapter 4. Interior architecture specializes in environments where people live, work and play in a diverse range of communities, as outlined earlier. Therefore, in a broad sense, due to the nature of the discipline, interior architecture is involved in working for the community at many levels.

There are numerous understandings of what community engagement means. Essentially, I am focusing on how interior architecture works 'with', not 'for' the

Fig. 2.3 Roles variously played by the different partners in a community initiative

community in order to address their values and needs. Community engagement needs to be thought of as a partnership. As such, Community engagement has the potential to 'identify the needs—or more important the assets [social capital]—of people and communities, and consequently to develop a range of options to consider' (Hendler-Ross and Hendler-Ross 2008, p. 13). These options may include environments, ways of working, transferable and discipline-specific skills and knowledge that are able to build the capacity of that particular community.

There are a growing number of examples of creative and built environment agencies seeking ways to enact social sustainability. Examples are illustrated in *The Great Good Place* by Ray Oldenburg (Oldenburg 1997), *Design for the Other 90%,* curated by Cynthia E. Smith (Smith 2007), *Expanding Architecture: Design as Activism*, edited by Bryan Bell and Katie Wakefield (Bell and Wakefield), and *Design Futuring: Sustainability Ethics and New Practice*, by Tony Fry (2009). Additional examples follow in the second and third sections of this book: *Social Justice and Interior Architecture* and *Cultural Heritage and Interior Architecture*. A number (but not all) of the examples indicate that they used Community Engagement principles either explicitly or implicitly in the way they facilitate social sustainability.

I reviewed a number of the aforementioned examples, and what emerged is that roles are being played that are not generally attributed to interior architecture or the profession's work with a community. These roles are being played interchangeably between designers engaging 'with' the community, being studied and acted on, and other partners working 'with' the community. The roles that emerged are: activist, educator, learner, designer, networker, strategist, fundraiser, builder (as in construction/manufacturing), leader, and champion. Figure 2.3 illustrates the roles that have been played by the various partners: the community, designers and other partners. In each initiative different partners can play different roles; these roles can be held equally by each of the partners, and/or the different partners may play only certain roles, depending on the context and the people involved. Through the lifetime of an initiative, partners may switch roles. These roles will be further explored in subsequent sections.

How Other Disciplines Engage with Community

The previous section provided an understanding of the concepts of community and engagement and their connection to social sustainability and interior architecture. This one examines how other fields tackle Community Engagement in order to address social sustainability.

Established Ways of Working 'for' and 'with' Community

Since the nineteenth century, medicine and law have established infrastructure within their professions to provide public services to those in need, including specialist education, internship and practice. Public health, for example, advocates for policy that extends community wellbeing. Additionally there are not-for-profit organizations such as the International Red Cross and Médecins Sans Frontières, where medical practitioners can volunteer their time for emergency and disaster relief and to assist with ongoing challenges in developing and developed countries. Law has a clear system for providing support in the public defence system (Fisher 2008).

Community development is often formally conducted by non-government organizations (NGOs), universities or government agencies to improve the social wellbeing of local, regional and, sometimes, national communities. Less formal efforts seek to empower individuals and groups of people by providing them with the skills they need to effect change in their own communities.

The social sciences have developed a large body of literature on research and practice in community engagement, with broad agreement on the basic concepts, principles and good practice approaches. Examples include: *Building Stronger Communities*, by Philip Hughes et al., and *Community Development and Partnerships: a Handbook for Building Community Partnerships*, by Florence Frank and Anne Smith (Hughes et al. 2007; Frank and Smith 2006). Table 2.1 provides a brief exploration of some of these concepts.

Medicine, law and the social sciences have developed a subset of their discipline that services community wellbeing. This raises some questions for the pedagogy of interior architecture: what might the specialization of 'Public Interior Architecture' look like? What would the education contain? There is a need to guide interior architecture students to look for collaborations in appropriate community service, design and built environment disciplines, and create meta-disciplines to develop vital and effective programs for social sustainability. Tony Fry explains that 'disciplinary thinking is exclusory', and therefore has a 'limited ability to comprehend and engage the complexity of unsustainabilty and the creation of sustainment' (Fry 2009, p. 55). He suggests that disciplines 'need bridging by a meta-discipline that facilitates an exchange of knowledge and dialogue based on a common language of engagement, while also amassing collective knowledge in their own right' (Fry 2009, p. 55). This thinking is not the same as in multi-disciplines or inter-disciplines.

Table 2.1 Community engagement related concepts. (Table adapted from DSE 2005, p. 15)

Concept	Explanation
Active listening (see Fig. 2.2)	Active listening is where an individual confirms that they have heard and understood by paraphrasing the information back to the speaker. It can be applied in many situations involving the engagement of others, including the facilitation and consultation processes used in Community Engagement. The 'talking wall' illustrates an oral and visual method
Appreciative inquiry and Assets based approach (see Case studies 2, 5)	Appreciative inquiry rejects the more traditional 'problem-focused' approach and instead seeks to identify what is working well, and links to the assets-based approach to community building. As an engagement approach it aims to encourage imagination, innovation and flexibility with stakeholder groups and build upon the positives which already exist, for example, collecting skills within the community, visioning a sustainable future. For example, The Asset-Based Community Development Institute at Northwestern University is part of a movement that considers local assets as the primary building blocks of sustainable community development
Community profiling (see Case studies 1, 2, 5)	Community profiling, as a scoping tool, is a means to achieve an increased understanding of the diversity of the community. Its purpose is to ensure inclusiveness and therefore a better engagement process and outcome. In participatory profiling the community is actively involved in the research, resulting in the community having greater participation in determining an appropriate course of action
Community strengthening (see Case studies 1–5)	The basic premise of community strengthening is that valuable knowledge and ideas are readily available within communities, and the role of others engaging is to develop mechanisms for sharing this knowledge. Community strengthening helps to mobilize community skills, expand networks, harness energy and resources and apply them in ways that achieve collaborative and positive social change
Systems thinking (see Case studies 1–5)	A systems approach encourages the exploration of the relationships between social, environmental and economic interactions This approach resists breaking a problem into its component parts for detailed examination. By examining the links and interrelationships of the whole system, patterns and themes emerge which offer insights and new meaning to the initial problem In a community engagement context, encouraging a diversity of views can lead to a new understanding of the situation and the identification of opportunities for action that may not have otherwise occurred (Bawden 1995, cited in DSE 2005, p. 15)

Emerging Ways of Working: Creating with Community

In the last section, communities of practice with established frameworks to research, sustain and engage with communities were briefly reviewed. Much is to be gained through investigating their methods, when developing interior architecture practices. In this section emerging models of community engagement are explored.

Interior architecture sits within a community of practice, defined here as the creative and built environment community. This includes allied disciplines in the built environment: architecture, engineering, furniture design, graphic design, industrial design, landscape architecture, textile and material design, urban design, rural and remote design, construction management, planning, as well as visual arts and craft, writing, film, and the performing arts. Within this diverse community of practice, there is a movement defined here as 'design activism', spawning research and action to harness shared knowledge that will develop socially sustainable communities. Over the last 10 years this has occurred in both developed and developing countries, in a range of situations resulting in transformation.

Creative and built environment case studies are herewith selected from literature and my own practice to illustrate a range of ways in which interior architecture may engage with communities. The case studies are examined through the lens of community engagement principles, in two aspects: ways to engage, and the benefits of engaging, associated with the purpose and outcome.

Case Study 1: Design Corps: Community Service Program and Design Network

Established in 1991 in the US, Design Corps' mission is to 'create positive change in communities by providing architecture and planning services. [The] vision is realized when people are involved in the decisions that shape their lives, including the built environment' (Design Corps 2011). Design Corps explicitly uses community engagement principles as 'the' way of working to achieve social sustainability and is one example of a number of design activism organizations globally. They are proactive, for they seek communities in need, invite them to engage, and set up a network to build and disseminate knowledge to others. Design Corps' objective is to provide affordable architectural services to those currently underserved by traditional practices. They have focused on two communities in need in the US—migrant workers and struggling farm communities.

Those engaging are: small rural communities, low-income families, farmers, farming communities and migrant workers, recent architecture and planning graduates, and students. Community members and Design Corps interns act in all the following roles (dependent on the project they are engaged in): activist, educator, learner, designer, networker, strategist, fundraiser, builder (as in construction), leader, and champion.

The ways of engaging include methods of community visioning and organizing, leadership to identify challenges, to pool needed resources, and to implement design responses through community involvement and participation (see Table 2.1). Design Corps provide technical assistance in planning, design, and grant writing.

An important extension of Design Corps' on-the-ground engagement is a network to build and disseminate information. This includes an annual conference, *Structures for Inclusion*, a website and related publications. These provide a space to share knowledge between interested communities, designers, funders, and policy makers.

For the community, the benefits of engaging are:

- Design and planning expertise, usually beyond their access, is provided, allowing the community to shape their physical environment and create positive change.
- Realization of projects, for example: the design, building and funding ($ 90,000) of a community garden and flea market; and the participatory design process, funding and construction of a $ 400,000 job-training centre.

For Design Corps and their interns, the benefits of engaging are:

- Interns learn by doing and developing core knowledge such as building practices and communication, as well as a range of skills and knowledge in community engagement, participatory design and vernacular design.
- Graduates are provided with a model, resources and networks to start their own design–build practices.
- They are enabled to refine strategies and build knowledge to disseminate in further community practice.

Case Study 2: Indigenous Community Volunteers (ICV): Asset-Based Community Building

In contrast with the last example, in this case study, set in Australia, the Aboriginal and Torres Strait Islander community develops and implements community projects, and invites others to engage with them. They call it 'reconciliation in action' (ICV n.d.). ICV was established in 2008, and illustrates how a community can be proactive in identifying their own projects in need of expertise, using community engagement to facilitate their social sustainability.

Those engaging are: Aboriginal and Torres Strait Islander communities, organizations, businesses, families and individuals; volunteers from all walks of life; and ICV. The community, ICV and the volunteers act variously in roles such as: activist, educator, learner, designer, networker, strategist, fundraiser, builder (as in construction), leader, and champion, depending on the project.

In engaging with community needs, ICV takes an asset-based, internally focused and relationship-driven approach. The 'asset' or strengths-based approach (rather than a deficit-focused one) has proved effective in community (and human) development, because it recognizes, celebrates and connects the existing capacities within a community (see Table 2.1). In this model, local strengths, knowledge, leadership and existing community relationships are mapped, connected and harnessed for the projects identified for development, with the objectives of improving the community's quality of life, health, social and economic wellbeing and inclusion within Australian society. ICV seeks out volunteers to work on community-owned and driven projects. The community identifies the skills needed, manages the project design and selection of volunteers to harness those skills, and runs the project in ways that work best for them.

For Aboriginal and Torres Strait Islander communities, the benefits of engaging are: realized initiatives, developed capacity and the benefits of needed external expertise. 'Strengths-based approaches create the transformation necessary to move from welfare to wellbeing' (ICV n.d.)

Case Study 3: XS Project: Community Building Through Capacity Building

This community building initiative in Indonesia, where XS equals excess, directly links environment, poverty, education and creative design solutions, and illustrates how innovation and design thinking can help build capacity and economic sustainability.

Those engaging are the 'trash picking community' in Jakarta, visual artist and environmentalist Ann Wizer, and the NGO formed from the initiative, XSProject. The community members' roles are: learner and builder (as in manufacturing), and Wizer's roles are: champion, activist, designer, educator, and learner.

The ways of engaging emerged from Ann Wizer's experimental art work using consumer waste as a direct way to address environmental damage and to encourage poverty reduction. In 2002 she started the XS Project by designing products from waste. The trash pickers sell their waste to the NGO (at higher than market price), and some work in the cottage industries that produce the designed items, providing two streams of income to the community. Additionally, the XS Project's profits help to educate children of the community.

For the trash picking community, the benefits of engaging are:

- education and training
- job creation
- new income opportunities for the very poor, mentally and physically challenged
- financial support for education for their children, such as supplies, fees and scholarships. (XSProject n.d.).

Case Study 4: Social Economic Environmental Design (SEED): Measurement Tool and Network

This example differs from the others presented so far in that it is a measurement tool, developed from 2005 to 2009 in the US by 'experts in the public interest design movement' to evaluate positive change brought about by community engagement initiatives in an identified community. It establishes measures for 'how design could respond to the triple bottom line of social, economic, and environmental justice', and allows evaluation of the success of a project. The latter is based on the objectives set by all the partners in an initiative, according to the agreed set of parameters (SEED n.d.).

> SEED maintains the belief that design can play a vital role in the most critical issues that face communities and individuals … SEED provides tools … that guide design professionals toward community-based engagement with design practice. These tools support a public-interest methodology that is increasingly recognized as an effective way to sustain the health and longevity of a place or a community as it develops over time. (SEED n.d.)

This system of evaluation is a valuable tool for interior architects wishing to establish the value of initiatives involving social sustainability. It is set up to address 'communities of location'; however, it may be adaptable to evaluate initiatives that work with communities of interest. There is the opportunity to sign up and use this system for interior architecture projects immediately.

Case Study 5: Lessons from Community Engagement Experiences

My commitment to a larger role for interior architecture in community engagement has come from my own practice, and from my observation of the practices of my colleagues where we respond to the needs of our local communities (in Perth, rural and remote Western Australia, and Australasia). Through a lifetime of exposure and practice I have built an engagement framework.

Embedded in my own experiences are attitudes and values that demonstrate that community engagement is not a job or task but rather, a way of operating. These were lessons learnt through relationships, beginning with my parents and continuing to the present.

Lesson One—Value the Individual

'Respect every person, no matter what their job is, who is doing the best job they can' was an often repeated lesson from my father. I took this to mean that every person has something to offer, and that people need to contribute, based on their capabilities and circumstances. My father specialized in public health in indigenous communities in rural and remote areas in Tanzania and Canada, my mother in midwifery. Their training and practice modelled a way of life and work that was reinforced for me by the institution called 4-H (Head Heart Hands Health). Working in rural Canada, 4-H's mission is to empower 'youth to reach their full potential, working and learning in partnership with caring adults' in order to prepare them 'to step up to the challenges in their community and the world' (4-H n.d).

Lesson Two—Strategic Thinking: Building Capacity, Leveraging Expertise

Ken Barlow was a consultant strategist who introduced to Marshall Cummings (now MC/IBI), the interiors practice I worked for, a system known as 'Project Planning Framework and Tools' in 1990; this provided the building blocks for the community engagement framework that evolved through my design practice, and later in my teaching and community initiatives described next. Barlow's framework differed significantly from the usual top-down, apprentice-like development of fresh graduates and intermediate staff, in that it assumed that less experienced practitioners come with know-how that is valuable to the project, and that the experienced practitioners' specialized know-how is required at specific times in the timeline of a project. The framework brings in such knowledge at the right time; the big picture project direction is made at the beginning by the team as a whole, and as the project progresses, knowledge and responsibility is moved down to less experienced members. The approaches introduced gave responsibility to the less experienced, mentored by senior staff, thereby accelerating capacity development.

Lesson Three—Knowledge is Transferable: Effectively Sharing it Accelerates Capacity

When I began academic life at Curtin University in 1993, by reframing these approaches in design thinking, capacity building and practice management to suit teaching and learning, I visualized our department as a practice, and students and academics as colleagues. As a group, we had a wealth of capabilities to share. Students were seen as individuals who had things to contribute, and were guided through various learning situations. Over time they became more confident and found a way to contribute that suited who they were.

The skill in building a community of practice is to foster the diverse knowledge of team members and encourage learning; then to recognize what skills and knowledge are required for the particular circumstance. Outside help can be called upon should required expertise not exist in the team assembled. From this experience I developed a pedagogical framework for community engagement.

Lesson Four—Motivation is The Key

In 2003 the designXchange, a collaborative of designer–makers, invited me to curate an exhibition of their furniture and objects (Lommerse 2004). We agreed that design development and prototype workshops would benefit the collective; the process was set up as a series of four intensive workshops that involved 25 interior architecture students.

I was not certain that the collective would adhere to the demanding work required to progress their prototypes during and between the workshops. Students are motivated in part by the rules of engagement in a university, where meeting deadlines is a requirement of the program, whereas designer–makers usually work more independently. I discovered these men were motivated, firstly by the desire to progress their work, and secondly, they had their sights set on an exhibition in the big smoke, Perth. Thirdly, each felt a commitment to the group and to meeting their objectives. These men spent a significant amount of time over an 8-month period on their developments, time where they sacrificed their incomes. It was at times very difficult for them. Individual and group rewards have to be valuable to the people involved for them to be motivated to work though these kinds of engagements—there has to be a hook.

Lesson Five—Strengths-Based Approaches Build Confidence

Following the designXchange exhibition, FORM (a not-for-profit organization which advocates for and develops creativity in Western Australia) invited me to work with them to research and develop the framework to bring together clusters of artists and designers. Over 2 years we worked collaboratively, through action research with pilot clusters, building the Designing Futures Cluster Development Program, conceptualized as asset-based, and structured by systems thinking (see Table 2.1). The research culminated in a written program and tools for various facilitators. The framework forms the basis of learning in the Midland Atelier, a creative arts hub in the Midland Railway Workshops (see Chap. 6). 'Led by renowned design mentors, Designing Futures delivers business skills and product development mentoring to specially selected "clusters" of Western Australian designers' (Designing Futures n.d). Each cluster focus is adapted to address different gaps in knowledge and learning, from initial product concepts and development, through to more advanced business mentoring or market placement.

Lesson Six—Community Engagement Framework can Align and Progress the Objectives of Diverse Partners

Perth is regarded as the most isolated city in the world. Designers are remote from markets and lack confidence in their standing in both a national and an international market. Therefore in 2006, Curtin University and FORM collaborated on a project to present West Australian artists and designers to an international audience at the Salone Internazionale del Mobile in Milan. *Riches of Isolation: Unearthing Western Australia's Creative Potential* (ROI) took 2 years from inception to completion and resulted in a travelling exhibition, a catalogue, public presentations and a public relations campaign (Lommerse et al. 2006).

Those directly engaged were the 23 invited exhibitors, 20 Curtin students, FORM and Curtin staff. Indirectly engaged were State arts and culture organizations, funding bodies, the design community, and the public, through publications, public talks, the exhibition and catalogue.

Ways of engaging: Using systems thinking, ROI was intended to build the confidence and capacity of the arts community, provide practising designer–makers and students with a place to act as learners and teachers, and facilitate deep and lasting connections between them. The aim of the facilitators was to teach by doing, illustrating that by working together, and imagining what seems to be unattainable, participants have the power to make things happen.

First, the community engagement framework was used with student–mentor teams to secure an invitation to the Salone Satellite, organize the funding and the exhibition entries. The Salone Satellite is the premier pavilion where the cream of international emerging design is exhibited. Highly sought after invitations are offered to only 20 institutions and 60 designers annually. Secondly, following the selection of exhibition pieces, exhibitors volunteered for various aspects of the project: photo shoot, publication research and writing, design and coordination of the travelling exhibition and finally, packing the works for airfreighting. Once in Milan, the exhibitors and curators set up, packed down and manned the exhibition.

The community engagement framework proved to be a powerful tool that can be used to align the objectives and agendas of a number of community arts organizations, along with students and graduates, in motivating collaborative work and community monetary support.

The benefits of engaging are:

- confidence—individuals' and community confidence in capability to participate at a global level
- communication—to new markets on a local, national and international level
- linkages and networks—to granting agencies, state policy developers, and the design community
- career pathways illuminated
- product sales
- invitations—to national and international exhibitions and symposiums, followed
- success in grants—to progress prototypes, followed (see Figs. 2.4 and 2.5)
- international experience and skills in event management, promotion and marketing design work, public relations and marketing, grant writing, project management, writing for publication.

Learning from Case Studies

My research has found evidence that a two-way growth is emerging between communities and the design professions. On the one hand, there is an intersection between communities who are looking for ways to address the problems facing them, using design as a tool. On the other, the creative and built environment disciplines have activated design thinking, systems thinking and community engagement as tools to address community issues.

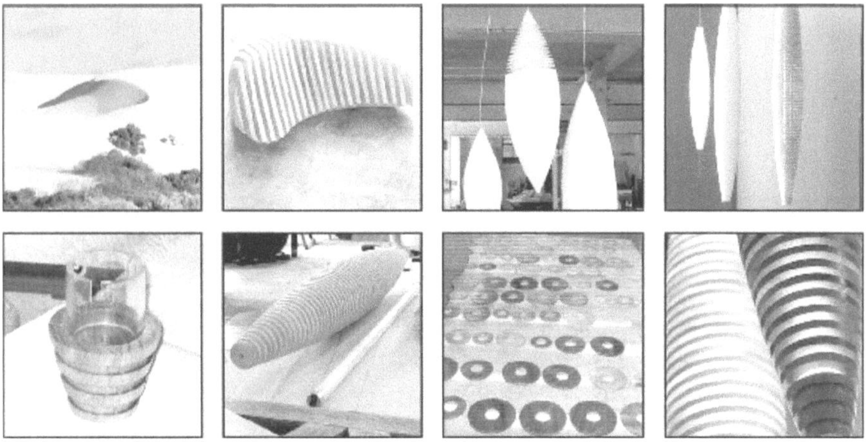

Fig. 2.4 Designing Futures facilitates capacity development through sharing of know-how, fostering relationships and building prototypes. Dune Pendant design and prototypes Malcolm Harris. (Photographers: 1, Malcolm Harris; 2–7: Marina Lommerse; and 8: Michael Chestnutt, with permission)

The case studies use design thinking and 'systems thinking' to explore the relationships between the social, environmental and economic issues facing the community. By examining the links and interrelationships of the whole system, patterns and themes emerge which throw light on the initial problem. In a community engagement context, encouraging a diversity of views can lead to a new understanding of the situation and the identification of opportunities for action that may not otherwise happen.

So what do capacity building projects, for instance, have to do with the core skills of interior architecture? For example, in ROI (see Lesson six), the design of the exhibition itself was an ephemeral space that was installed for 6 days in Milan and for 2 months in Perth. But this was only a part of where our skills were employed. The four curators/project managers (Marina Lommerse, Penelope Forlano, Rebbeca Eggleston and Kris Brankovic) could be described as a meta-disciplinary team; we combined our knowledge, strengths, and circle of influence to champion a project, harnessing the very different capabilities of the two organizations we represent—one a large university, the second a small but connected State arts and culture organization—together with the know-how of practitioners and students (Lommerse et al. 2006).

The more I learn and work with different areas in the design sector, the more the experience confirms the relevance of the inter-related principles underlying effective management, team building, teaching and learning. The process is about engaging people through good communication and respect for the individual, building teams with a range of capabilities and knowledge, and designing clear and achievable project objectives and structures—all community engagement principles. When I reflect on the lessons learned in Case study 5, I see how intuitively and pragmatically I used these principles, and so I conclude that they may be harnessed more in interior architecture practice, teaching, learning and research.

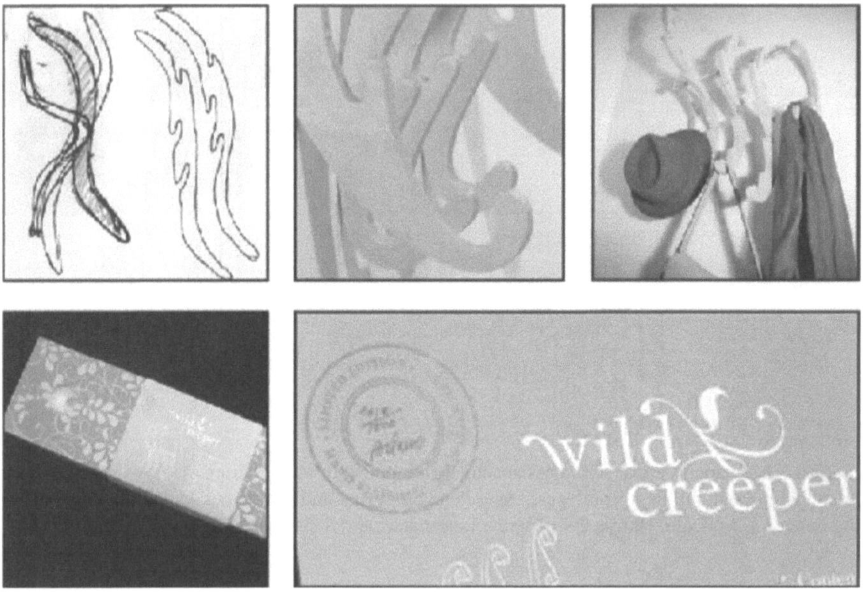

Fig. 2.5 Penelope Forlano's 'Wild Creeper': taking product from idea to prototype, to packaging and marketing, through her involvement in Designing Futures and ROI. (Photographs by Forlano, used with permission)

My core work as an interior architect deals with relationships. I and my team deal with the relationships of the people and the activities going on, in whatever environment we are working towards designing. Often these activities are ephemeral, and the environment needs to be designed for various activities and people. We, the design team, hone our skills in understanding complex situations, and the environments that facilitate people. The design of spaces that meet the needs of people and their activities is complex and multi-layered. But it is interior architects' attunement to people and their relationship to activity and space that makes us so valuable in the area of community engagement.

The processes we use as interior architects are complementary to community engagement. We gather information about the needs of a situation, imagine and make visible a variety of solutions, and take a solution from an idea into a realized form, harnessing the resources and meta-disciplinary knowledge required to make it a reality. Important, well-honed core skills that interior architects can offer when engaging with a community are project management strategies: the ability to pre-plan a project and its phases over a long period of time; the ability to understand what skills will be required and when on a project timeline; and strategies for dealing with hiccups in the schedule. Unexpected changes in direction can happen, and we have the capability to take on new information, remember the old information and strategize a way forward. Case study 5 illustrates how I have applied methods of practice to higher education structures and to my teaching.

A clear framework for how to get from the beginning to the end, and what role each will play in that picture is important. There needs to be a link between the objectives of the community initiative and the individual. There has got to be something in it for them; this seems an obvious observation. However, it is important. In the case studies described previously, the partners in the project had to engage individuals to participate directly by giving their time and energy and sometimes money, and other stakeholders to give their support in engaging with the initiative. When the going gets tough, and it does, individuals and organizations need this motivation in order to keep at it. This is called the 'hook'—it is what motivates them.

Integrating Community Engagement into Interior Architecture

I started this study looking for where the potential is for interior architecture to engage with communities. The case studies in this chapter, and in other chapters of the book, illustrate the power of community engagement to enable short term impact and long term transformation. The examples demonstrate that community engagement involves personal connection and values that are integral to the process. The literature, combined with personal experiences, offer a number of insights on how it can be extended. Currently, interior architecture is limited in the way it works with community; in this section I explore the possibilities.

The case studies presented show that effective engagement can lead to the identification of mechanisms for building a community's strength and its ability to join with other stakeholders in dealing with complex issues and change, thus affecting social sustainability. Community engagement can be used to build more resilient relationships with community, and some of the values that emerge are:

- engagement can grow a sense of identity within the community, unique to each project
- engagement can foster a sense of belonging to community and considerable benefits for the agents from working together on behalf of the community, creating a synergy and learning from each other
- there may be more ownership of solutions to current problems or building plans for the future, so that the community shares in decision-making and has a higher level of responsibility for creating that future
- individuals may become empowered and proactive with regard to issues that affect them
- communities are able to identify their priorities for themselves
- opportunities are created for a diversity of voices to be heard on issues which matter to people
- positive, planned change can be transformational for the community and individuals within it.

Community input can improve the quality of policy being developed, making it more practical and relevant.

In *Design Futuring*, Tony Fry proposes that in order to ensure the future (of humanity), humans need to rethink the design processes we use, as the current design paradigms are defuturing (that is, everything we create destroys something; the current rate of creation and destruction means that humans have a finite future) (Fry 2009). Considering the critical concerns about social sustainability, and the ability for community engagement to facilitate it, we need to consider how to integrate community engagement and interior architecture.

In order to integrate a new way of working into a community of practice, the structures and practices that inform it need to be considered. These structures are education, professional bodies, practices and the research and dissemination of knowledge. Therefore, if interior architecture wants to embed community engagement in a way that is effective for the community, these structures must be integrated into the mandates and policies of the professional organizations and make a place for research, education and internship to feed into the practices. For example, a specialization in public interior architecture might guide students to courses in the university that deal with community, and internships in community design would prepare students for this kind of career track. In the creative and built environment disciplines, post-graduate degrees of this kind exist, such as The Design Academy Eindhoven's Masters Course (Social Design Research Program), and the Escola Tecnica Superior d'Arquitectura's Masters of International Cooperation (The Design Academy Eindhoven n.d.; Schachter 2008). And Curtin University's Bachelor degree in Interior Architecture focuses one third of the program on socially sustainable design (Curtin University n.d.). But there is a limited offering globally.

Earlier, assumptions about the narrow practice of community engagement in interior architecture were outlined. I suggest that if interior architects could see the wider potential and range of applications, they may embrace the concepts. The research has shown that there is a broader relevance for it. Interior architecture using community engagement as a mode of practice can creative positive changes in communities of interest, communities of location, and communities of need. In addition, Community engagement can be used effectively to promote social justice and cultural heritage. By defining and expanding the concept, we can open up career paths and practice areas and play a vital part in influencing policy development for communities.

The very nature of the work of interior architecture has the potential to make invaluable contributions in community engagement. For example interior architects deal all the time with:

- the people–environment interface
- very complex issues (using systems thinking), and integrating these issues throughout the collection and establishment of the brief or program, design phases, construction and on-site work. We synergistically work to integrate the program complexities, and often have to revisit and reconstruct as challenges occur, and as we deal with ephemeral spaces that require future-focused thinking

- interior architecture has the potential to connect communities to their culture through facilitating cultural expression in their interior spaces; through consultation we can open up the discussion and seek to engage the whole community in the development of place. Our ability to turn idea and concept into visual expressions and propose real spaces allows clear communication with the community.

Taking Action

Action to foster social sustainability is explicitly supported by the International Federation of Interior Architecture/Interior Design (IFI) and World Interiors; however, there is limited evidence that education, internship or interior practices have structures that provide methods, case studies or direction in the area (see Chap. 4). Interior architecture has got to start somewhere, so here are some suggested actions:

- Name interior architecture actions, and make them visible using social networking.
- Get involved in established projects through NGOs (see ICV, Aid to Artisans—individuals can register online to work with these organizations).
- Engage with established and effective networks that are actively building and disseminating information; be discerning, and don't reinvent the wheel. Inform these networks of the know-how interior architecture can bring.
- Be proactive; seek communities that have a need for design thinking. Size is not important, so you can get involved at a micro or macro level.
- Make other fields working with community aware of what interior architecture can offer.
- Reflect, evaluate effectiveness and disseminate the work of interior architecture.
- Use established evaluation tools (see SEED).
- Learn, seek tools and training, seek expertise from established community organizations in the region (see DSE).
- Educators, review your curriculum and pedagogy. Are your methods of learning or methods of working with clients and communities oriented only towards the status quo? (See Eindhoven, Escola Tecnica Superior d'Arquitectura; Curtin University Interior Architecture Program).
- Universities can look to create clusters with various disciplines including interior architecture by challenging entrenched traditions to create a collaborative paradigm (social sciences, health, film, public health etc.) (see Imagining America n.d.; Qureshi 2008).
- Work locally. Advantages of working locally have been espoused by a number of design activists. Working locally offers a number of advantages, as over time a series of linked initiatives can foster transformation, attract champions and know-how, and access an ongoing involvement over a period of years. It builds stronger links and trust with the community, so that the community begins to see themselves in their stories, and builds confidence in their voices and networks (see University of South Australia, Student Construction Program; and Fig. 2.6, below. D. Morris, the Coordinator of the Student Construction Program at the

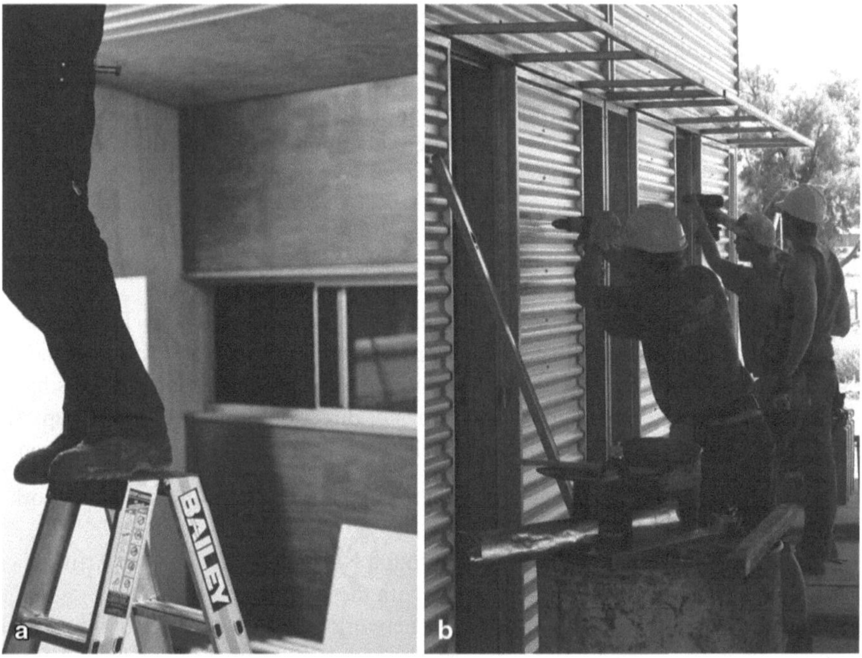

Fig. 2.6 **a** and **b** Mimili Community, South Australia: the University of South Australia, School of Art, Architecture and Design. Student Construction Program has been working with Aboriginal communities since 1993. (Photographers: *Left*, C. J. Taylor; *Right*: Hendrix Chan, used with permission)

School of Art Architecture + Design, has been running consultation and design build programs with communities since 1993).

- Engage with innovative ways to use technology to facilitate participation, for example the use of social networks.

Conclusion

In a substantive essay in the catalogue for the exhibition, *Permeable Boundaries: Interiors—a Field of Possibilities*, Dianne Smith and I argued that interior architecture has untapped potential, unique to our discipline, outside of our understood practices, and beyond the narrow public understanding of what our discipline does, and how it can contribute to the good of the public (Lommerse and Smith 2010). The interior architecture community, using our core knowledge, can enrich and open up opportunities for other communities.

It may be perceived that I am trying to establish disciplinary boundaries around community engagement and interior architecture, despite theories that support meta-disciplinary cooperation. This is not my intention. However, I maintain that interior architects can be proactive. Rather than just responding to needs identified by clients, we can look for places where our capabilities can contribute. Roberta Feldman wrote that being proactive is the act of 'leaving the office, engaging a community and seeking a need for design in the community, rather than passively waiting for clients to come to them' (Feldman 2008, p. 15). She argues that to be proactive our community of practice needs to broaden our scope to include:

> any intentional, creative action—formal, programmatic, fiscal, functional, physical, social, political, or aesthetic—that achieves lasting positive change. We need to re-imagine our practices in order to empower ourselves to make the contributions we believe [interior] architects can and should make. (Feldman 2008, p. 15)

I hope that by looking at 'community' and 'engagement' in this chapter from multiple contemporary understandings, it opens up previously unrealized opportunities for interior architecture to participate in creative work for the benefit of communities.

References

Bullock, A., & Trombley, S. (Eds.). (2000). *The new Fontana dictionary of modern thought*. London: Harper Collins.

Badanes, S. (2008). Building consensus in design/build studios. In B. Bell & K. Wakeford (Eds.), *Expanding architecture: Design as activism* (pp. 248–255). New York: Metropolis Books.

Bell, B. (2008). Expanding design towards greater relevance. In B. Bell & K. Wakeford (Eds.), *Expanding architecture: Design as activism* (pp. 15–16). New York: Metropolis Books.

Curtin University. (n.d.). http://www.curtin.edu.au/course_overview/undergraduate/interior-architecture. Accessed 14 Feb 2014.

Cohen, A.P. (1985). *The symbolic construction of community.* London: Tavistock.

Community Renewal International. (n.d.). http://www.communityrenewal.us/.

Design Corps. (2011). http://www.designcorps.org/about.

Designing Futures. (n.d). http://www.form.net.au/industry-development/designing-futures.

Feldman, R. (2008). Out of the box: Design innovations in manufactured housing. In B. Bell & K. Wakeford (Eds.), *Expanding architecture: Design as activism* (pp. 208–219). New York: Metropolis Books.

Fisher, T. (2008). Public-interest Architecture: A needed and inevitable change. In B. Bell & K. Wakeford (Eds.), *Expanding architecture: Design as activism* (pp. 8–13). New York: Metropolis Books.

Frank, F. & Smith, A. (2006) *Community development and partnerships: A handbook for building community partnerships*. Perth: Curtin University.

Fry, T. (2009). *Design futuring: Sustainability, ethics, and new practice*. Sydney: University of New South Wales Press.

Hendler-Ross, A. & Hendler-Ross, S. (2008). Designing with an asset based approach. In B. Bell & K. Wakeford (Eds.), *Expanding architecture: Design as activism* (pp. 123–131). New York: Metropolis Books.

Henri, F., & Pudelko, B. (2003). Understanding and analysing activity and learning in virtual communities. *Journal of Computer Assisted Learning, 19,* 474–487.

Hughes, P., Black, A., Kaldor, P., Bellamy, J., & Castle, K. (2007). *Building stronger communities*. New South Wales: UNSW Press.

Imagining America. (n.d.). Artists and scholars in public life. http://imaginingamerica.org/about/. Accessed 2 Dec 2013.
Indigenous Community Volunteers (ICV). (2011). http://www.icv.com.au/. Accessed 2 Dec 2013.
Little, A. (2007). Community. In B. Galligan & W. Roberts. (Eds.). *The Oxford companion to Australian politics* (pp. 120–122). Oxford University Press, Melbourne.
Lommerse, M. (2004). (Curator and Ed.). [Exhibition and catalouge]. *designXchange*. Perth: Craftwest [now FORM]
Lommerse, M. (2009, July). Facilitating cultural transformation: Redefining indigenous identity through architecture. Cultural crossroads: Proceedings of the 26th International SAHANZ Conference (p. 1–22). https://www.academia.edu/198402/Facilitating_Cultural_Transformation_Redefining_Indigenous_Identity_through_Architecture. Accessed 2 Dec 2013.
Lommerse, M., & Smith, D. (2010). (Curators and Eds.) [Exhibition and catalogue] *Permeable boundaries: Interiors—a field of possibilities*. Perth: John Curtin Gallery.
Lommerse, M., Forlano, P., Eggleston, R., & Brankovic, K. (2006, April). (Curators). [Exhibition and Catalogue]. *Riches of isolation: Unearthing Western Australia's creative potential*. Perth: FORM. https://www.academia.edu/636243/Riches_of_Isolation_unearthing_the_creative_potential_of_Western_Australia. Accessed 2 Dec 2013.
Oldenburg, R. (1997). *The great good place*. Philadelphia: Da Carpo Press.
Qureshi, S. (2008). Enhancing family and community through interdisciplinary design. In B. Bell & K. Wakeford (Eds.). *Expanding architecture: Design as activism* (pp. 263–273). New York: Metropolis Books.
Scott, J., & Marshall, G. (Eds.). (2009). Community. In *A dictionary of sociology*. Oxford University Press. Oxford reference online. http://www.oxfordreference.com.dbgw.lis.curtin.edu.au/views/ENTRY.html?subview=Main&entry=t88.e337. Accessed 2 Dec 2013.
Schachter, A. (2008). Teaching cooperation. In B. Bell & K. Wakeford (Eds.), *Expanding architecture: Design as activism* (pp. 255–263). New York: Metropolis Books.
Social Economic Environmental Design (SEED). (n.d.). http://seednetwork.org/. Accessed 2 Dec 2013.
Smith, C. E. (2007). (Curator). [Exhibition and catalogue]. Design for the other 90 %. New York. Cooper-Hewitt, National Design Museum, Smithsonian Institution.
Smith, M.K. (2001). Community. *The encyclopedia of informal education*. http://www.infed.org/mobi/community.htm. Accessed 14 Feb 2014.
The Department of Sustainability and Environment. (2005) *Effective engagement: Building relationships with community and other stakeholders, Book 1*: *An introduction to engagement*, Version 3. Melbourne: Victorian Government Department of Sustainability and Environment. http://www.dse.vic.gov.au/__data/assets/pdf_file/0019/105823/Book_1_-_An_Introduction_to_Engagement.pdf. Accessed 2 Dec 2013.
The Design Academy Eindhoven. (n.d.). http://www.designacademy.nl/MasterCourse.aspx.
Tropman, J. E., John, E., Ehrlich, J. E., & Rothman, J. (2006). *Tactics and techniques of community intervention*. Belmont: Wadsworth Publishing.
XSProject. (n.d.). http://www.xsprojectgroup.com/. Accessed 2 Dec 2013.
4-H. (n.d.). http://www.4-h.org/about/. Accessed 2 Dec 2013.

Chapter 3
More Than Bricks and Mortar: Lessons in Design and World Citizenship

Kelley Beaverford

Abstract Designers, like all world citizens, must move towards creating more socially sustainable solutions. To respond to the demands of working in a global society, the Faculty of Architecture at the University of Manitoba, Canada offers a course called Service Learning in the Global Community. In a six-week experience, students engage in a design-build project with a community overseas. While learning about design and construction, participants gain skills in cross-cultural collaboration and world citizenship. Offered in rural communities in Turkey, Uganda and Ghana, the course impacts on the lives of a diverse group of participants, including young designers. This chapter reflects on the objectives, the themes of engagement, the role of reflection, and the challenges and limitations of this course.

Keywords Service learning · Interior architecture · Architecture · Community engagement · Intercultural learning · Co-design · Turkey · Uganda · Ghana · Social sustainability

Introduction

Globalization has created an increasingly interconnected world, placing new demands on the design professions. Designers, like all world citizens, must move towards creating more socially sustainable solutions. Good design has the potential to promote the wellbeing of all populations, provide equitable solutions, and create a better future for an interconnected world. To respond to the demands of working in a global society, the Faculty of Architecture at the University of Manitoba, Canada offers a course called Service Learning in the Global Community. In a six-week experience, students engage in a design–build project with a community overseas. While learning about design and construction, participants gain skills in cross-cultural collaboration and world citizenship. Offered in rural communities in Turkey, Uganda and Ghana, the course impacts on the lives of a diverse group of participants—including young designers. This chapter reflects on the objectives, the themes of engagement, the role of reflection, and the challenges and limitations

K. Beaverford (✉)
Department of Interior Design, Faculty of Architecture, University of Manitoba, Canada
e-mail: Kelley.Beaverford@ad.umanitoba.ca

D. Smith et al. (eds.), *Perspectives on Social Sustainability and Interior Architecture*, DOI 10.1007/978-981-4585-39-2_3, © Springer Science+Business Media Singapore 2014

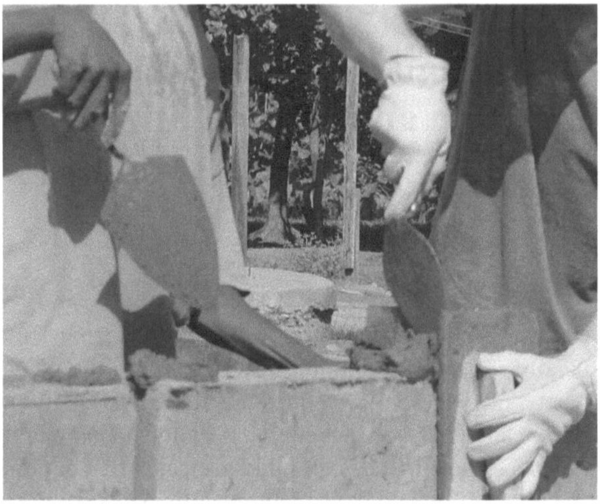

Fig. 3.1 Collaboration, photographer Kelley Beaverford

of this course. Examples are drawn from our most recent project, the Saint Anne's Boarding House in Damongo, Northern Ghana (Figs. 3.1 and 3.2).

Learning Objectives

Service Learning in the Global Community is primarily concerned with strengthening the intercultural dimensions of design education. A common goal between a community and university group serves as a catalyst for cultural exchange; in this case, the goal was to construct a kitchen, dining and study hall for Saint Anne's School.

Students work with community leaders, builders and residents, while serving as junior designers and apprentices on a worksite. Displacement intensifies the experience by encouraging participants to see beyond their own social realities while making connections between design, culture, climate, economics and construction. The disparities serve as a reminder that designers are rarely experts in a new or unfamiliar reality; therefore, they must suspend their assumptions to become effective in a community project. The main goals are to integrate experiences and provide opportunities to reflect within the context of design education.

The learning objectives can be divided into two categories: international knowledge and intercultural competencies. International knowledge is primarily concerned with declarative knowledge, procedural knowledge, and culture-specific knowledge of the host country, partner organizations, and project (Vulpe et al. 2001). Pre-departure sessions offered information about the country as well as exposure to people who have lived and worked in Ghana. The background of the partner organizations and the project was also covered prior to departure. The students learned of a need to improve access to education for girls in Northern Ghana. Some of the common

Fig. 3.2 Saint Anne's Boarding School project in progress, photographer Kelley Beaverford

obstacles between girls and education in this region are large domestic workloads, pressure to marry and tuition fees (Shepherd 2010).

Canadian students learned of the systemic factors that link education to development, along with the political, economic and climatic implications informing the design and construction of buildings in the region. The knowledge-based objectives for this course are to:

- learn about people, materials and construction through experience
- provide challenges through which students can see beyond their own space, time and culture
- integrate existing and new theories, methodologies, technical skills and practices into the process of making design decisions
- promote understanding of the complex interdependence of global economic, political and cultural forces affecting the built environment.

Intercultural competencies are mainly concerned with how we approach given situations in cross-cultural contexts. Culturally effective people learn to appreciate differences in approaches and how to adapt to new situations. For instance, Canadian participants in the Saint Anne's Boarding House project noted cultural differences that affected design and management decisions in Ghana. According to student journals, examples observed on the worksite include: the Ghanaians' greater respect for authority and tradition, higher tolerance for uncertainty on the worksite, and lower priority on time and schedules. By working with locals over an extended period of time, students began to understand and respect the differences between cultures. Seminar discussions, readings and integrated experiences provided opportunities to reflect on how increased intercultural competencies can promote good practice and design. This was especially relevant for those preparing to practice internationally or in multicultural societies. The objectives for 'intercultural competency' within the Service Learning in the Global Community course are to:

- enhance cross-cultural communication skills
- engage in interdependent learning between all participants through engagement

- develop effective strategies and design solutions for addressing challenges in appropriate, respectful, and constructive ways
- challenge perspectives on social problems of others who are in different social or cultural groups than themselves.

Themes of Engagement: Interdependence, Reciprocity and Mutual Benefit

Service Learning in the Global Community builds on the three themes of interdependence, reciprocity and mutual benefit. The broad concepts are described in this section and their application in this study and the implications for future work are discussed.

Interdependence

There are typically three partners in a service-learning experience: the community, the students and the university (Canadian Alliance for Community Service Learning 2007). An inclusive partnership with the non-profit organizations was formed through direct action as co-educators and managers. The long-term commitment to the community played a pivotal role in them identifying a project that would address actual needs. The Saint Anne's Boarding House project had two non-profit organizations representing the community—the Catholic Diocese of Damongo and Tools for Schools Africa. Both groups have demonstrated success in the region and have earned the support of the community. Other community partnerships were also formed. Most notable is the relationship with the local contractor and builders, who engaged in daily interaction with the students and instructors as mentors, co-workers and friends.

The students and faculty members are equally significant partners. An interview process helped identify suitable students from architecture, interior design, landscape architecture and environmental design. Although we encourage students with a strong academic record to apply, the selection process is primarily concerned with attracting a diverse group of students who represent a variety of educational and life experiences. Similarly, the faculty members bring experience in four or more disciplines, including interior design, architecture, photography and graphic design. All of the faculty members who have participated in the course have taught and practised design in cross-cultural settings. And, as a representative of the university and course instructor, I agreed to serve as the project manager responsible for coordinating the construction of the building and the course. Other members of the University of Manitoba community provided technical, administrative and fundraising support.

Although involved in an informal capacity, teachers, school children and other community members also played vital roles in the success of the course.

Further success was achieved through collaborative and interdependent relationships. In contrast to traditional design studios, Service Learning in the Global Community does not promote design as a solitary endeavour. The design–build pedagogy adds a visceral connection to the collaborative process of producing the built environment. Design and construction solutions are resolved on-site within a distinct social, economic and cultural context. The goal is to impel students to make respectful and well-informed decisions by broadening their exposure to interdependent relationships.

Reciprocity

Service Learning in the Global Community applies a principle of reciprocity, promoting a mutual exchange of action and benefits among partners. Above all, we do not consider the community as a classroom in which students try out their skills and ideas, nor is the course about charity. Service learning—in its most basic form—is about learning with other people. Every participant functions as both a teacher and a learner (Canadian Alliance for Community Service Learning, 2007). In this way, a strong collective can be formed and meaningful dialogue can occur. As researcher Novella Keith explains, 'Service learning must involve more than contractual relationships, calling for dialogue not only as an exchange of ideas but an encounter between fellow human beings' (Keith 2005, p. 15). This encounter—or mutual exchange—motivates participants to acquire the knowledge and competencies described as part of this course. Moreover, a high level of interaction raises the prospect of arriving at mutually beneficial and empowering solutions in the built environment.

Diversity of the participants serves as a key resource in teaching and learning. Sociologist Jui-shan Chang refers to this idea as a 'transcultural wisdom bank'; Chang suggests that by taking advantage of diversity in a class, we can generate a comparative perspective in the curriculum, while increasing intellectual depth and personal engagement (Chang 2006, p. 370). In Service Learning in the Global Community, the wisdom bank is trans-cultural, interdisciplinary and intergenerational. As a result, the Saint Anne's Boarding House project was informed by various design professionals, trades people, development workers, educators and students representing a wide range of life experiences. Exposure to different perspectives enables participants to see beyond their own social realities and thereby gain further understanding of their own cultural and disciplinary perspectives.

Building on the notion of trans-cultural exchange, participants learned lessons in vernacular design from daily interactions with community partners, combined with a six-week stay in Damongo. The construction of the boarding house kitchen provided opportunities to learn of the close relationship between culture and design. For example, unlike food preparation spaces in Canada, the Ghanaian kitchen required vast open spaces for physical activities such as pounding *fufu* (cassava) or stirring *banku* (maize flour). In the same manner, the exuberant use of colour

Fig. 3.3 Craftsman demonstrating wood work; photographer Kelley Beaverford

combinations on building exteriors prompted explanations of availability of paint, maintenance and local associations with building typologies. The Canadian students were intrigued by the realities of working in a tropical climate, as these impinge on the design, construction and maintenance of a building. Comparative analysis of the differing contexts of building practices offers a lens for understanding how and why we create space, while encouraging participants to develop adaptable ways of thinking about design.

Cooperation across demographic borders can also encourage greater understanding of differences in construction practices. The students developed respect for the handcrafted solutions employed by Ghanaians and learned to make foundations, masonry blocks, window frames and trusses on-site and without the convenience of power tools (Figs. 3.3 and 3.4). In journal entries and seminar discussions it was evident that Canadian students preferred the tradition of handcrafting work—used among Ghanaian builders—to practices prevalent in their own culture. The partners noted that Ghanaian participants had a higher tolerance for uncertainty with regard to finished products and schedules, and that they placed greater emphasis on collectivism than their Canadian counterparts. This was specifically demonstrated in the cohesion of the builders and their devotion to the contractor on the worksite, and their tendency to live with extended families. It is interesting to note that these differences between West African and Canadian values, as observed in this study, complemented the ones documented in Geert Hofstede's work on cultural dimensions (Hofstede, n.d.).

The students observed the environmental impact of building a structure in rural Ghana. The comparisons being shown between Canadian and Ghanaian building practices brought concepts to the foreground that scholars such as Tony Fry have proposed: that western concepts of sustainable design often focus on maintaining practices that were not necessary or sustainable in the first place (Fry 2009). In Damongo, sustainable building practices—apparently driven by need rather than desire—evoked many reflective journal entries. Students wrote about re-use of materials on-site, preference for local materials, adaption of tools, and natural ventilation found in vernacular building design and construction (Fig. 3.5). Others speculated on how 'leapfrog' technology could be further enhanced if combined with long-term, culturally-informed development strategies (Kennedy 2007). Daily

Fig. 3.4 Craftsman demonstrating how to apply plaster; photographer Kelley Beaverford

Fig. 3.5 Re-use of materials, students straighten nails; photographer Kelley Beaverford

interaction with builders facilitated valuable discussions about the responsibility of designers to create environments that do not threaten the health and wealth of the global community. Through reciprocal learning, the designers and builders learnt of the responsibility to avoid inequitable, myopic solutions during the course.

Mutual Benefit

In Service Learning in the Global Community projects, the partners join forces to understand and address community needs. In addition to the common wealth of the community, each partner may have individual motivations for participation. Students may be motivated by the high level of need of the community and a desire to learn about cross-cultural design. Similarly, the community may be motivated by a desire to complete a project, matched with an interest in addressing the needs of the students. The efforts of participants are most effective when the projects are mutually beneficial for everyone involved.

The mutual outcomes often include short-term impact and long-term transformational change. Collaboration with community partners informed the students of the skills and knowledge required to work in a rural West African community. This engagement also offered a marginalized population the opportunity to contribute to the development of international design education and architectural practices.

Although the builders functioned like mentors, they were also students of cross-cultural collaboration. The Canadian students shared their perspectives on design, construction and development. They also demonstrated safer workplace practices, such as wearing steel-toe boots and hard hats, both practices embraced by the Ghanaian builders. Further, by combining the expertise and resources of the Catholic Diocese of Damongo, Tools for Schools Africa, and the University of Manitoba, we were able to strengthen the community's capacity to provide safe, affordable housing for 60 girls in the region. The short- and long-term goals were to support a sustainable program, in numerous stages of development, thereby encouraging girls to complete secondary and further education. The strategy included constructing the boarding house with local builders, contributing technical and financial aid, and creating opportunities for residents of the boarding house to meet women undertaking university programs.

Although international service learning can provide significant contributions to communities, the benefits to students of the course are equally impressive. Experiential learning creates opportunities to apply theoretical knowledge to real-life situations, increasing the meaningfulness of course work in professional programs. Critical thinking is encouraged through discussion and exposure to systemic relationships that influence design; such as social realities, culture, construction, economy, and the environment. Community engagement can provide students a competitive advantage to succeed in a global economy by promoting enhanced international knowledge and competencies through a collaborative process. Accordingly, Service Learning in the Global Community offers opportunities for students to participate in transformative learning experiences while learning about the connections between world citizenship and design.

Universities also stand to benefit from international community engagement. The projects can support research, service and teaching initiatives for academics committed to strengthening the social responsibilities of higher education and design. Hands-on experience provides unparalleled junctures for learning intercultural competencies and the impact of culture on design. Further, the design–build approach helps to shed light on strengths and weaknesses of the technical aspects of existing programs. Areas for research include—but are not limited to—project-related issues such as social and environmental sustainability, development, international practice, building science, and the internationalization of curriculum. The institution also benefits from a greater connection to the global community, helping educators improve curriculum as well as attract and retain students and faculty concerned with world citizenship.

Reflection

Reflection, in its various forms, plays a key role in a service-learning experience. It provides opportunities for participants to understand how engagement impacts their perception of the community as well as their roles as professional and global citizens (Canadian Alliance for Community Service Learning, n.d.).The tools for reflection in this course include formal and informal discussions, a journal assignment, photography, design and hands-on construction.

In Ghana, the reflection between the students and community partners was typically informal, emerging in conversation over meals or on the construction site. Many students have maintained their friendships with Ghanaian builders after their return to Canada. More formal reflective interaction has transpired between the organisers of the course, starting almost 1 year before the build and continuing after the completion of the course. In anticipation of future projects, representatives from the Catholic Diocese, Tools for Schools Africa and the University continue to evaluate the strengths and weakness of the project.

The students reflected on the experience through design, readings, student-led seminars and a structured journal assignment. The seminars provided a supportive and respectful environment for different perspectives to be heard. Each student was required to submit 15 journal entries at the end of the course. The assignment asked that the entries make connections between experiences in the community, required readings, and concepts and theories covered in previous courses at home. Journal entry topics included: cross-cultural design, development, design–build, studio education, building science, working abroad, sustainability and community engagement. Students were encouraged to derive meaning and knowledge from experiences that occurred before, during and after the service-learning experience.

Challenges and Limitations

A successful service-learning course presents many challenges and requires a high level of commitment from all partners. The involvement of Tools for Schools Africa and the Catholic Diocese of Damongo preceded the University's contribution. Their ongoing commitment to the community—specifically to Saint Anne's Boarding House—should be noted. We received a warm welcome in the town of Damongo due to their exemplary work and strong, nurtured relationships with the community. The depth of their understanding of the community served as a guide to the university participants in achieving meaningful, appropriate and respectful engagement opportunities. The faculty members had to demonstrate due diligence and patience in their efforts to find partners of such high calibre.

Hosting foreign students and faculty members can present many challenges for the community partners. Although the university group tried to be considerate guests, the partners undoubtedly felt pressure to ensure that we had an enjoyable and productive stay. This was evident in upgrades to the guesthouse and meals, along with provision of daily transportation to the site. The time constraints of the course also presented challenges to the local contractor and builders. As Canadians tend to work with a linear task orientation of time, the Ghanaian hosts struggled to make their typically more flexible ways meet the expectations of the Canadian participants. This became apparent in changes to their traditional construction schedules and the addition of more men on the construction site towards the end of the course. Moreover, assuming the role of co-educators can create obstacles in the way of meeting deadlines. And while the students worked extremely hard, they often lacked the expertise required to complete tasks as efficiently as experienced builders.

The students faced challenges similar to those of international volunteers. Many students identified with the cons of international volunteering described by Michael Palmer (Palmer 2002). The experience of being a visible minority—though sometimes beneficial—exhausted some students as they felt that they were constantly on show. Participants also commented on feelings of frustration, loneliness, and exhaustion due to high temperatures, lack of personal space, cultural differences and an intense work schedule sustained over a six-week period. Likewise, the persistence of inevitable travel health problems such as diarrhoea, heat rashes and insect bites can be hard on even the most experienced travellers. The students were concurrently expected to deal with the complex intellectual and physical demands of the course. Perhaps it is not a surprise that many of the students affectionately described the course as 'extreme' in their journal entries.

From a faculty member's point of view, some of the greatest challenges lie in the pre-departure work. Finding suitable relationships between community partners and the course must be dealt with well in advance. From there, it becomes a matter of coordination between partners, a site visit, preliminary design, budgeting and fundraising. Recruiting strong candidates for the course, and pre-departure training also takes time and careful consideration. Offering the course—as an instructor, negotiator, guest and project manager—involves a significantly higher level of responsibility, energy and time commitment than traditional teaching assignments require

Conclusion

Service Learning in the Global Community begins to support the goal of enhancing design education through world citizenship. It starts by recognizing a need to expand the professions, to become more inclusive in the way we work and more responsive to the communities we serve. To do this, we must consider the gaps between what we have today and what we need to know in order to create more socially sustainable environments. While international community engagement presents many challenges, ultimately the substantial benefits make it worth the effort. With this in mind, community engagement, when informed by the themes of interdependence, reciprocity and mutual benefit, can provide designers with lessons on more than bricks and mortar.

Acknowledgements The author would like to thank Tools for Schools Africa, the Catholic Diocese of Damongo, the community of Damongo, Claudius Bangfo and his crew, Professor Leland Hill, Professor Michael Hersrud, Kayla Schlosser, Desiree McKay, Lindsay Greschuk, Michelle

Tustin, Amanda Jones, Jordy Craddock, Christine Plett, Marc Mainville, Marianne Moquin, Architects Without Borders Canada, and the University of Manitoba for their contributions to this project.

References

Canadian Alliance for Community Service Learning. (2007). *Principles of good practice for community service-learning.* http://www.communityservicelearning.ca/en/documents/InformationSheets-PrinciplesofGoodPracticeinCSL-2007.pdf.

Canadian Alliance for Community Service Learning. (n.d.) *Reflection tools for faculty.* http://www.communityservicelearning.ca/en/resources_information_sheets.htm.

Chang, J. (2006).Transcultural wisdom bank in the classroom: Making cultural diversity a key resource in teaching and learning. *Journal of International Education, 10*(4), 369–377.

Fry, T. (2009). *Design futuring: Sustainability, ethics, and new practice.* Oxford: Berg.

Hofstede, G. itim international. (n.d.) *Cultural dimensions.* http://geert-hofstede.com/hofstede_dimensions.php?culture1=14&culture2=99#compare.

Keith, N. Z. (2005). Community service in the face of globalization: Rethinking theory and practice. *Michigan Journal of Community Service Learning, 11*(2), 5–24.

Kennedy, S. (2007). Leapfrog: Design strategies for global innovation. In C. Smith (Ed.), *Design for the other 90 %* (Exhibition catalogue) (pp. 60–65). New York: Cooper-Hewitt National Design Museum, Smithsonian Organization.

Palmer, M. (2002). On the pros and cons of volunteering abroad, *Development in practice, 12*(3), 637–643.

Shepherd, J. (2010, Feb. 23). What is a girl worth? *The Guardian.* http://www.guardian.co.uk/education/2010/feb/23/ghana-education-girls-attitudes-resources. Accessed 24 April 2010.

Vulpe, T., Kealey, D., Protheroe, D., & MacDonald, D. (2001). A profile of the interculturally effective person. Centre for Intercultural Learning, Canadian Foreign Service Institute, Canada.

Part II
Social Justice and Interior Architecture

Within the scope of this book, the intersection between social justice and interior architecture is explored as a critical aspect of social sustainability. The aim is to identify, capture, describe or critique the issues and practices that are emerging during the current decade, with the aim of fostering both action and future research within the discipline.

There is a rising responsibility for those in more privileged positions to address issues and events that impact adversely on the quality of life of sub-groups and populations during the twenty-first century. This trend gives emphasis to the need to address or counteract inequality and the infringement of basic human rights. Such issues can be described within the domain of design and social justice.

The trends toward inequality in society are not inevitable and can be modified by sound public policy based on a robust version of the concept of 'equal worth'. Measures which improve the economic status of the least well off, increase employment, reduce inequality and 'civilize' the workplace are likely to produce significant improvements for all of us. Conversely, passivity, selfishness and indifference to the fate of our fellow citizens is certain to lead to unnecessary and significant social dislocation. It's all a question of what we are prepared to do (Lawrence 2002).

Reference

Lawrence, C. (2002, September). What is social justice? Trilogy of conversations about social justice (Part 1). Address delivered from Hollis Lecture Theatre, Curtin University, Bentley. http://www.safecom.org.au/social-justice.htm. Accessed 24 Nov 2013.

Chapter 4
Interiors can Address Social Justice: Fact or Fiction?

Dianne Smith

Abstract Interior architectural designs can influence the quality of a person's day-to-day life. However, there is an apparent absence of such professionals practising within projects and policy development that relate to environments for members of our community who are considered to be without, or who are marginalized in either local or international contexts. Therefore issues of Social Justice are raised and the concept defined. People may be disadvantaged due to conditions beyond their immediate control—whether these conditions be medical, political, cultural or geographical. The key characteristics of interior architecture are discussed to demonstrate how practitioners and educators could contribute to improving the quality of lives of such people. As background to this proposition, a study to ascertain whether interior architecture is involved in socially sustainable projects and debates was undertaken in 2010. The question posed was: 'what is the nature of the profession's involvement and impact?' This was answered in three ways. Firstly, the positioning statements of major professional design bodies were reviewed. Secondly, interior-based projects and the people involved were sought. Finally, literature from other fields was sourced to gain insights. The findings and their implications for the future directions of interior architecture are discussed in relation to the field of social sustainability. In response to the study's outcomes, a snapshot is constructed which enables roles and strategies for interior architecture and associated designers to be formulated.

Keywords Social Justice · Interior architecture · Design · Marginalized · Profession · Social sustainability

Introduction

A contradiction exists that has sparked my curiosity for some time, and it is this matter that I wish to interrogate and make explicit, in order to promote discussion and foster action. What is this contradiction? On one side is the inherent power that

---

D. Smith (✉)
Department of Architecture and Interior Architecture, Curtin University, Perth, Australia
e-mail: dianne.smith@curtin.edu.au

D. Smith et al. (eds.), *Perspectives on Social Sustainability and Interior Architecture*,
DOI 10.1007/978-981-4585-39-2_4, © Springer Science+Business Media Singapore 2014

interior architecture has to impact upon the quality of people's day-to-day lives.[1] On the other, is the apparent absence of the profession within projects and policy development relating to environments for members of our community, who are considered to be without or who are marginalized in either local or international contexts. I argue that it is critical to conceptualize interior architecture in terms of its key characteristics in order to envisage the role that practitioners and educators can play in impacting on the quality of lives of people who are disadvantaged due to conditions beyond their immediate control—whether these conditions be medical, political, cultural or geographical.

In 2010, I undertook a study to ascertain whether interior architecture is involved in socially sustainable projects and debates. The question posed was: 'what is the nature of the profession's involvement and impact?' This was answered in three ways. Firstly, the positioning statements of major design bodies were reviewed. Secondly, interior-based projects and the people involved were sought. Finally, literature from other fields was sourced to gain insights. This chapter reports on the findings of this study and reflects on their implications for the future directions interior architecture may take in the field of social sustainability. Before outlining what is happening or missing in this space, the concept of social justice in relation to design will be defined. The survey undertaken and its findings are then described. In response to the study's outcomes, a snapshot is constructed which enables roles and strategies for interior architecture and associated designers to be formulated.

What is Social Justice in this Context?

Social justice is not about charity. It is a belief that all human beings are entitled to basic rights, to be treated with dignity and to have a sense of identity within their day-to-day life. In association, it is the right for all people to be treated equitably in regard to resources and conditions so that they may, at a minimum, experience a basic standard of living that provides security, shelter and comfort. The enablement of people's basic rights for facilities and dignity must be couched in terms of equity of access. Lack of education or finance, for example, may prevent some people from taking an opportunity that is accessible to others. As a consequence, although to the outsider there is opportunity for all, in reality access is compromised for such individuals.

The terms 'disadvantaged' and 'marginalized' recognize that within our global and local societies, many do not have equal access to a range of facilities and services—and, in some cases, to a basic quality of life. For example, those with mental health issues may find themselves in substandard boarding houses, indigenous communities may not have access to adequate water or sanitation or culturally sensitive

[1] The term 'interior architect' is used interchangeably with 'interior designer' in this chapter. It should be noted that, unlike many practitioners globally, practitioners in Australia refer to themselves as 'interior designers', due to the Architects Act, which protects the word 'architect'.

buildings, those living in settlements (such as refugee camps or shanty towns) may need to scavenge materials to create shelters or dwellings and in some cases one- or two-room homes may need to be sublet to assist survival. In addition, floods, famine, war and economic circumstances can mean that many others can never break the poverty cycle. Payment for the upgrade of facilities such as sewerage, water or refrigeration is impossible, and homes are adapted with what is at hand to suit the circumstances. Even today, there are people with physical disabilities who do not have equal access to interior spaces, simply because of where they reside or work. In addition, the need to be situated within culturally or sub-culturally appropriate milieux is a significant one.

It is important to clarify that I am not referring to people who live differently to others and who are content in the way that they exist. Rather, I am referring to those people who, if they had the resources, would enjoy the benefits of alternative facilities and services in relation to basic human needs and rights. As Abraham Maslow (1954) described, there are core needs associated with being human that need to be met; 'physiological needs'; 'safety needs'; 'belonging needs'; 'esteem needs'; 'need to know and understand'; 'need for aesthetic beauty'; 'self-actualization'; and 'transcendence' (as cited in Rowan 1998, pp. 81, 82, 90; Changing Minds 2010).

Therefore, the discussion crosses continents and diverse groups of individuals. For example, the increase in homeless populations in developed countries highlights that such issues are part of a global trend and are not just characteristic of societies challenged by the extreme circumstances suggested above (Kothari 2007). 'One in 200 Australians are homeless without safe, secure or affordable housing' in 2010, according to Homelessness Australia (2010 n.d.). The Department of Families, Housing, Community Services and Indigenous Affairs (FaHCSIA) reported in 2009 that there were 201,000 homeless in France, 500,000 in Germany, and 113,590 people in England who had a priority need. Canada's homeless growth rate was described in a special report to the Human Rights Council (2009, p. 16) as 'unprecedented since the Second World War'.

As a consequence, design in relation to mobile, temporary, flexible, crisis or long-term homes—or more simply to the provision of shelter—is linked to local as well as cultural and political agendas, which may vary from country to country or suburb to suburb.

How can interior architect participate in the debate and in programs that address social equity and Social Justice for all members of society?

Because of the nature of the profession's core business, interior architects can address these issues by using their creative talent and processes to envisage new ways of tackling old problems. Such consideration will result in strategies to develop environments in which to live, work and play, where the quality of our and others' lives is enhanced, not lessened. It should be noted that interior architects, while they deal with the intimate scale of surface, texture, furniture and furnishings, also at a deeper level, deal with the quality of space and the relationship between people and place. Their work is critical in the facilitation or hindrance of meaningful relationships for individuals and groups within the spaces they inhabit; this broader dimension will be discussed more fully later.

What is the Explicit Position on Social Justice of the Professional Interior Architecture Bodies?

The peak international body of the interior architecture and interior design profession, the International Federation of Interior Architecture/Interior Design (IFI), explicitly supports socially responsible design and states:

> … it is a mandate to use Interior Architecture/Design as a tool to improve the quality of life and uphold human dignity for all. Socially Responsible Design can help alleviate a range of problems caused by poverty, illness, lack of basic shelter, or badly designed physical environments. IFI advocates supporting the human dignity of all people through Interior Design by changing the perception that design remains the domain of an élite segment of the world's population; and encouraging professional designers to use their best abilities as creators and innovators to solve social and environmental problems. (IFI n.d.)

Part of IFI's activities is the Pro Vitae program. Its goals are to 'make social contributions globally using the talents of Interior Architects/Designers' and to support projects which are 'voluntary philanthropic endeavours of a high professional level and in the field of Social Design; covering issues such as Universal Design, Barrier-free Design, Sustainable Design, Design for All and Design for Emergency' (IFI n.d.).

Although IFI has socially responsible design as its mandate, this is not explicitly evident for other national professional bodies, although they are affiliated. The websites for many professional bodies drawn from a range of countries do not indicate they are engaging in the debate or undertaking programs that align with such agendas. For example, the websites sourced included the Malaysian Society of Interior Designers (MSID), Chinese Society of Interior Designers (CSID), Korean Society of Interior Designers (KOSID), Indonesia Society of Interior Designers (HDII), New Zealand Society of Interior Designers (NZSID), Indian Institute of Interior Designers (IIID), Philippine Institute of Interior Designers (PIID), Interior Design Association, Hong Kong (HKIDA), Interior Design Confederation, Singapore (IDCS), Japan Interior Designers Association (JID), Thailand Interior Designers Association (TIDA), and Japan Society of Commercial Space Designers (JCD).

Others, however, do give some indication. The American Society of Interior Designers (ASID) acknowledges contributions through the annual ASID Design for Humanity Award. In a study mutually funded with Interior Designers of Canada (IDC) et al., the knowledge base of interior architecture was compiled in 2005. Of the six areas identified, (Human Environment Needs, Design, Products and Materials, Interior Construction, Codes and Regulations, Professional Practice, Communication) human environment needs sits most comfortably in association with humanitarian agendas. The authors state that this area includes the contribution that interior architecture 'makes to the public's life, health, safety, and welfare', and draws from literature covering not only everyday environments, but also specialized environments such as residences for people with Alzheimer's Disease (Martin and Guerin 2005, p. 52).

In contrast to the mandates outlined by the IFI, the International Interior Design Association (IIDA) does not specifically outline its position on such projects. Policy Statement 1.3 on sustainability, however, briefly alludes to a promotion of social responsibility, stating that: 'IIDA seeks to promote the awareness and knowledge of interior design strategies that reduce negative impact on our natural environment and improve the health and wellbeing of all people', and 'we need to design with socially responsible materials and construction methods that benefit everyone that touches that project' (IIDA n.d.).

These imply that social responsibility is a positive consequence of achieving environmentally sustainable design.

The European Council of Interior Architects (a member of IFI), have had their ECIA model Code of Conduct and Professional Ethics for Interior Architects adopted by all the organization's members since 2005, including Belgium Association of Interior Designers, Association of Italian Interior Designers, Bund Deutscher Innen Architekten (BDIA), Bund Osterreichischer Innenarchitekten (BOIA), Beroepsvereniging Nederlandse Interieurarchitecten (BNI), Association of Interior Designers/Interior Architects of Madrid, Interior Designer's Union Estonia, Association of Furniture and Interior Architects Finland, Norwegian Architects and Interior Designer's Association, Fédération Nationale des Syndicats d'Architectes d'Interieur (FN SAI), Interior Designers in Ireland (IDI), Association of Polish Artists, Finnish Association of Interior Architects, Swedish Association of Architects and Association Suisse des Architectes d' Intérieur.

In section "What is Social Justice in this Context?" of the model, it states the ECIA's responsibility to society thus:

> The Interior Architect accepts a professional obligation to further the social, cultural and environmental standards of the community. He takes into account the consequences of his professional activity to the health, safety and well-being of all those who may reasonably be expected to use or enjoy the product of his work. (ECIA n.d.)

The Union also has standards for the level of professional education outlined in the 'European Charter of Interior Architecture Training 2007'. In all these programs, information about and understanding of environmental ethics and the role of sustainability in the practice of interior architecture are recorded. However, the Union, from information provided on websites, does not publicly cite any projects related to Social Justice previously or currently being undertaken by themselves or by one of their 16 members. Further, none of the organizations publicly cite a Social Justice project or provide additional statements about their position towards Social Justice in relation to interior architecture. If projects do exist, they would potentially complement the code of conduct outlined by the ECIA in 2005.

In Australasia in 2007, the Queensland Chapter of the Design Institute of Australia (DIA) initiated a working group to develop a socially responsible subgroup to integrate designing and designers in pro bono projects for marginalized people or communities. This subsequently developed in 2010 into DIA: Action's national initiative.[2] The Designers Institute of New Zealand's (DINZ) website mentions little on their stance toward Social Justice or related projects being undertaken by members. However, according to their Code of Ethics 3.1, 'members shall endeavour at all times to further enhance the design standards of the community' (Designers Institute of New Zealand 2003, p. 1).

[2] The author was involved in this process as part of the Queensland chapter, recorded in board minutes, but not published; the group is no longer active.

The South African Institute of the Interior Design Professions (iid) is explicit in its involvement in socially responsible work. As the public interface, its website refers to numerous instances of their strong position on socially responsible design and the 'IFI Think Tanks' taking place across regional grouping that include Europe, Asia, Oceania, India, Singapore, Africa, Middle East, North and South America. 'Each think tank will focus on an in-depth conversation which will explore the value, relevance, identity and responsibility of the interior's field' (iid n.d.). In a review of the AGM, the newly appointed President stated that a key forward-looking objective of the organization is to 'promote design awareness in disadvantaged communities' (iid insight 2010, p. 8). Regarding their involvement in socially responsible projects, the review states that the Institute is currently working closely with the Greenside Design Centre. The Centre facilitates tertiary students to undertake a great number of social design projects. As part of this initiative, the Institute and its members act as collaborators, sourcing projects, assessing the viability of student proposals and arranging sponsorship for those which can be carried out.

The United States and Canadian Interior Designer Educators' Council, IDEC, has a socially responsible design subgroup. It has a Community Service Award, which:

> … recognises and honors significant contribution of community service by an individual or group at a national, regional, and/or local level associated with the discipline of Interior Design. Service can include a variety of activities including but not limited to: making a difference to a community or special needs group, service-learning projects that build the awareness and value of Interior Design, efforts in the advancement of and dedication to the promotion of IDEC, effort in the advancement of Interior Design education or design scholarship, or an activity demonstrating significant community engagement. (IDEC 2010a)

The Flip Side: Engaging with Interior Architecture

The above section has reviewed the public face of the professional bodies in interior architecture, and discussed whether and how Social Justice and socially responsible design are highlighted by these bodies as significant issues. The other side of the equation to be discussed is the recognition or acknowledgement of these issues by non-government organizations (NGOs) or government organizations such as AUSAID (Australia's aid agency) or UNESCO (United Nations Education, Scientific and Cultural Organization). Three examples are mentioned here to demonstrate the potential for involvement.

Firstly, among their many areas of relief work, the Australian Red Cross is working nationally and internationally on emergency shelter and transitional housing construction projects. These are simple and primarily involve local labourers with the help of engineers, architects and construction workers. The involvement of interior architects is not mentioned (Australian Red Cross n.d.).

Secondly, the Salvation Army works across a vast range of areas world wide, including aged care, community and family support, counselling, court and prison services, disability services, domestic violence, drugs and alcohol, employment, family tracing, homelessness, problem gambling, suicide prevention and bereavement, youth issues and pastoral care. To support these aims, the Salvation Army is engaged in a number of infrastructure projects and advocates the benefit of good socially responsible design. In their response to the 'Regulation and Growth of the Not-for-Profit Housing Sector' discussion paper, the Salvation Army outlines the following principle as applicable to their service provision:

> Tenants should have access to the range of services which are integrated into a housing assistance model. The Salvation Army has had significant service experience through its long term, transitional and crisis-short term housing provision in addressing these issues of support. (The Salvation Army 2010, p. 4)

Thirdly, UNESCO's Division of Cultural Expressions and Creative Industries is committed to promoting the socio-cultural and economic role played by crafts in society through programs such as DESIGN 21 and TRIBUTE 21. Working currently in Paris, the Philippines, Saudi Arabia and India, the sector's aim is to prove to affiliated authorities that the artisanal sector deserves priority in national development plans (UNESCO n.d.). Projects outlined through the UNESCO website have the potential for involvement of interior architecture, but no reference is made to this area of design through the organization's publications, although reference is made to landscape architects, graphic designers, architects and urban planners.

From this overview, which is based on the public face of the profession, it is evident that the attention to socially responsible design, Social Justice and social equity through or in association with interior architecture is random or limited. Although the potential is high and an intention is evident, it appears that such work is yet to be considered as part of core professional interior practice. Instead the focus of this area is largely associated with the design of health-oriented facilities and universal design.

In addition, other designers avoid the issue, as they advocate or consider that designing for 'the user' will capture all possible user groups through good design processes. This unfortunately is not necessarily the case, for a large percentage of people do not have access to interior environments informed by or created by designers—particularly interior architects—to enhance the quality of the environment, and in association, their everyday life. Due to lack of time, lack of funds, the scale of need or the absence of a discrete client, a number of potential projects are never conceptualized and/or attended to by the interior architecture profession.

Mapping Contemporary Projects Involving Interior Architecture Internationally

In order to ascertain whether the conclusions drawn from the review of major design bodies are true, the second stage of the exploratory study was undertaken. The aims of this part of the project were to:

- ascertain if interior architecture practitioners were involved with Social Justice issues
- analyse the nature of the profession's presence and, if possible, its impact
- generate a summary of the modes of operating

- profile the potential of interior architecture to inform and challenge the profession and others.

Literature—including academic papers and government and industry reports—was sourced. Projects involving interiors and the names of interior architects undertaking projects were identified through the literature and by web searches. A surprising lack of projects and contributors, who could be described as interiors-based, meant that a snowball approach needed to be adopted in order to locate the hidden or less publicized players or projects. This research method relied on referrals from individuals and resources to form an ever-increasing network of contacts and data. As a slow but rewarding process, it revealed a cross-section of people who are working in this area for universities, NGOs and government, as well as individual practitioners and interior companies. However, unfortunately, many of these contacts led to dead ends or non-interior projects or connections.

The importance of the study is that it revealed a number of ways or modes in which interior architects can be involved. The fifth to eighth chapters in this volume present the work of some of these people, who each raise issues that need to be considered as we move into the middle of the twenty-first century. The remainder of this chapter will summarize the modes of operating and the potential impact in the context of current practice. In particular the findings are reassessed in light of the strengths of interior architecture as a profession and the potential it holds to participate in this field; from this appraisal, a way of positioning the interior architecture profession is proposed.

Modes of Contribution

The examples identified can be described under the following modes:

- hands-on participation in interior-focused projects as interior architects
- hands-on participation in collaborative projects as interior architects
- educational projects involving interior architecture students
- interior-focused projects involving other professions but not involving interior architects
- installations
- advocacy and endorsement
- research studies and conferences
- labour
- financial support
- exhibitions.

As information was at times limited, it is acknowledged that there may be other examples and modes that could be added in the future, and additions to this overview are welcomed. Each mode is defined through this study as follows, and examples are identified. Further examples are also provided in Table 4.1 and the relevant mode noted.

1. Hands-on participation in interior-focused projects as interior architects
   Projects that deal with the design of an interior space in order to address how a person function, interprets and/or experiences the space and associated activities, and constructs it subsequent meaning. Core skills pertaining to the discipline are drawn upon to execute the task.
   Example: No explicit examples were identified
2. Hands-on participation in collaborative projects as interior architects
   Projects that involve a number of disciplines working together to address an issue or execute a building project. The interior architect's core skills pertaining to the discipline are drawn upon to execute the task.
   Example: Ngari School Construction project, Solomon Islands (Ngari School Construction project n.d.)
   Erin Foley, an Australian interior architect, contributed to the Ngari School Construction project, Solomon Islands, during 2009 with Emergency Architects. The school was previously housed in a tent. She worked with the community to create woven window panels and walls for the new school buildings.
3. Educational projects involving interior architecture students
   Projects incorporated into the curriculum of colleges or universities where social issues and community responsibility are highlighted.

   – Hypothetical: The focus is to raise awareness in young designers and to build their skills as well as to develop processes. They come to understand that design can make a difference across a range of projects, including those that have a restricted budget or cater for a diverse range of people, such as those who are less fortunate or non-mainstream. Students may interact with the hypothetical end-users.
   – Actual: Although the focus is the same as the hypothetical, these projects have immediate outcomes by generating ideas and possibilities for actual clients, and/or tangible outcomes that can be or are built.

   There are many academics who discuss such issues, but there are few who actually undertake such projects. However, examples are available of both actual and hypothetical projects, and it may be inferred from the growing number of educational projects that community awareness and professional involvement may increase in the future.
   Example: Western Carolina University Service Learning Program
   This interior architecture program, run by Jane Nichols, is an example of using service learning and civic engagement to demonstrate the value of interior design to and for the community. Students worked with tribal representatives, members of the community and organizations to complete projects associated with a Cherokee Wellness Spa and the Harris Regional Hospital. By applying knowledge of colour and surface treatments, students also provided 'face lifts' for non-profit agencies such as women's shelters and child-abuse screening centres. The Appalachian Farm and Preserve 'aimed to re-establish traditional farming, craft and survival skills, on which today's heritage tourism economy is now based'. Following visits to the farmstead, the students:

> … designed the buildings to reflect the architectural vernacular of late 19th century Appalachia, created floor plans and selected sustainable exterior and interior building materials for the school buildings. They also developed the unique programming and designed all interior spaces to accommodate the activities in each space. (NCU Center for Service Learning n.d.)

4. Interior-focused projects involving other professions but not involving interior architects
   Projects where other disciplines or consultants take on the role of the interior architect.
   Examples: Rodney Harber and Angela Buckland (Chap. 6) and Paul Memmott (Chap. 7).
   Although from different continents, Rodney Harber (South African architect and academic) and Paul Memmott (Australian anthropologist, architect and academic) are renowned for their work dealing with disadvantaged indigenous populations where poverty and health are major issues. Both discuss issues associated with their and others' interior work.
5. Installations
   Projects that promote engagement with community, and/or activation of spaces or exploration of issues and ideas. Some may involve advocacy. Such projects are not reliant on the core skills of interior architecture practice. However, they may draw upon communication, organizational and facilitation skills as well as creative thinking.
   Example: Butterfly Works, Netherlands
   Butterfly Works describes itself as an 'inspiration-driven social solution and design studio'. Although based in the Netherlands, its projects have included consumer social project design and distribution, and the development of digital and mobile education materials and learning environments, while always emphasizing 'the centrality of the end user in the design process' (Butterfly Works n.d.). Butterfly Works designs objects, provides raw materials, provides microcredit to purchase materials, and trains women, thereby helping to counteract local women's loss of core skills due to a lack of opportunity. They work on small insertions such as canopies from palm leaves and weavings as well as larger architectural installations. The Cultural Centre and Eco-Lodge, in South Western Sahara, responds to the aim 'to build a Cultural Centre focusing on preserving both cultural traditions and environment in combination with empowering the local community' (Butterfly Works n.d.).
6. Advocacy and endorsement
   Professional bodies and individuals may act as advocates to assist in issues that are associated with people who have a lesser voice. This can occur through design guidelines, policy formation, and/or through endorsement of activities and functions.
   Example: Jill Pable and Lisa Waxman (Chapter 5).
   Jill Pable and Lisa Waxman, from the United States, outline how their work on homelessness and affordable housing illustrates the connection between design, policy and society.

7. Research studies and conferences
   Design practices and individual practitioners may undertake research projects with the intention of understanding in-depth issues relating to this domain. Issues may also be explored and discussed through conferences and seminars as a form of professional development, consciousness raising and/or policy development.
   Example: Art & Design for Social Justice Symposium, Florida
   The annual symposium's agenda is to critique how 'the tools and inherent abilities within the areas of art and design can be utilized in addressing issues confronting less advantaged groups within our local communities, states, regions or world' (Waxman 2007), and involves the disciplines of architecture, interior architecture, art education, fine art, dance, theatre, music, film, and social services (Art and Design for Social justice Symposium n.d.; IDEC 2010c).
8. Labour
   Individuals may give of their time and skills to carry out tasks including working bees, selecting finishes, lobbying suppliers and the like. These may or may not be linked to their core professional skills.
   Example: Kyabra Project, DIA Design: Action, Queensland, Australia (Wright 2010).
   This project involves the client, builder, consultants and academics working with a range of design discipline practitioners to create residential development that addresses universal design principles. Under the auspices of the Design Institute of Australia, many practitioners, including interior architects, provide their skills and knowledge.
9. Financial support
   Professional bodies, design practices and individual practitioners may contribute funds to support 'good causes' as part of their way of operating.
   Example: American Society of Interior Designers (ASID) Foundation
   The American Society of Interior Designers (ASID) Foundation website states that for all shopping done with a 'participating merchant, a percentage of what you spend will be returned to the ASID Foundation' (American Society of Interior Designers n.d.).
10. Exhibitions
    Exhibitions serve to explore issues associated with Social Justice, profile design products or processes that may assist in improving life quality, and/or display the works of disadvantaged communities to help develop their potential. Exhibitions involve interior architecture through the design and curation of the exhibition as well as the content.
    Example: Design for the Other 90 %, New York
    Cynthia E. Smith, curator of the New York's Cooper Hewitt National Design Museum, oversaw the development of the exhibition, Design for the Other 90 % (Cooper-Hewitt National Design Museum n.d.). The emphasis was on product design, although some spatial design examples were included such as the Day Labour Station, designed by architects, Public Architecture, San Francisco.
    The following table cites further examples of interior architecture in relation to design for social responsibility. This list is not exhaustive but does serve to demonstrate the diversity in this relationship. It is hoped that other interior-focused

Table 4.1 Further examples of design for social responsibility

Project	Description	Mode
Women's health services, CODA-STUDIO, Australia	Women's Health Services (WHS) was the first not-for-profit health centre established in Western Australia. CODA and Ian Dewar Architects were appointed by the East Perth Redevelopment Authority to design new premises incorporating two heritage-listed worker's cottages and a warehouse. Co-founder of CODA Emma Williamson trained in Interior Architecture. CODA also has a pro bono program to 'facilitate voluntary work to assist community-based groups to pursue or complete projects for their organisation' (CODA-STUDIO n.d.)	2
A community studio, Drexel University Philadelphia, USA, 2008	Students at Drexel University redesigned a printing factory building into a facility located in a low socio-economic area with 'high unemployment, substandard housing, increasing crime rates, [and] inadequate education'. The aim was to provide a centre where community groups could worship, learn and socialize, and the adopted process involved the community members as part of a participatory-design studio (Drexel University 2008)	3
Urban Studio, University of North Carolina, Greensboro, USA 2007	Robert Michel runs The Urban Studio, where students have undertaken two design–build community projects involving retired couples, or abused mothers and their children. The process involved the home owners, faculty and city representatives including those from Housing and Community Development, YMCA, Department of Justice's Office of Violence Against Women, and Technical Community College (Urban Studio 2007)	3
Design Corps, Bus Shelters, North Carolina, USA 2005	A Western North Carolina design–build program involved Architecture and Interior Architecture disciplines to design bus stop shelters. Community involvement was integral to the project's success. Various organizations collaborated (including Housing Services, Community Foundation, Department of Transit and Technical College) (Design Corps 2009)	3
University of South Australia, cross-disciplinary project for Indigenous communities, 1993–ongoing	Founded by architect David Morris, students are engaged in cross-disciplinary teams designing and building for particular communities in regional and remote Australia. Students can select from three elective study modes: design consultation and design, prefabrication and on-site construction. The building systems are developed in the university workshop and transported to site. The designs are developed in response to community needs (University of South Australia n.d.)	3, 8

Table 4.1 (continued)

Project	Description	Mode
Make a Difference Project, 2005–ongoing	This is an initiative, inspired by Bryan Bell, founder of Design Corps, of Interior Design Educator's Council (IDEC) Social Responsibility Network. In 2009, this project's third iteration involved 50 projects completed by 142 students who took up the challenge to design for positive change through a real-life problem in the community (IDEC 2010b)	3
Creative Practice as Research, Curtin University, 2009–ongoing	Interior Architecture senior students generate a hypothetical proposal to assist a select marginalized user group for their final design studio. Some students extend this philosophy into their Dissertation projects with outcomes that may inform policy, design thinking and/or future action. Example topics include: a culturally-sensitive methodology for foreign designers who can't be on location (Helen Mather); rethinking residential care for people with psychotic illnesses (Gaia Sebastiana); integration of a hybrid model to generate culturally sensitive housing models for Indigenous populations (Vaelei Walkden-Brown)	3
Kids' Guernica International Peace Mural Project	The project involved workshops for children in different places across the world to help foster peace consciousness through art (Kids' Guernica n.d.)	5
Artist Projects, Copenhagen, 2010	Marianne Frandsen discusses issues associated with the role of the artist and the interior architect and how the former's attitudes could inform the latter in Chap. 8	5
Structures for Inclusion Conference, 2010–ongoing	Architecture for Humanity conference held in affiliation with Design Corps. Three streams ran—design activism, development as change agent for the good, and politics and policy. The theme was: 'How can we build on the success of the Green design movement in addressing critical social and economic issues through design' (Architecture for Humanity 2010)	7
Designing for People with Cognitive Impairment, QUT and Curtin University, Australia, 2000–ongoing	Projects to investigate the links between the built environment and how people with cognitive environments cope in their everyday lives. These include studies into people with brain injury, autism, and dementia; and focuses on how the interior impacts on learning and quality of life (Smith 2009, pp. 220–227)	7
Colmeia Community Project, Sao Jose da Lapa, Brazil, 1997	Founded by psychologist Virginia de Macedo Torres, the Colmeia Community Project helps the residents of the low-income community by providing free classes in art, crafts, languages, IT, music and dance. The Colmeia Community Project building was designed and supported by Juliana Torres, an Interior Designer. (Torres, personal communication 2010)	8, 9

Table 4.1 (continued)

Project	Description	Mode
EDRA Conference 2009: The Ethical Implications of Sustainability in Interior Design Intensive Session	The conference website states: '… the objective of this intensive session is to discuss the current state of interior design and the ethical implications of sustainability'. Practice-related presentations include: Social and economic sustainability: creating new models of cohabitation in the existing urban home by Shelly Duff and Carolyn Gibbs; Reduction of operating energy as a design imperative for Interior Designers by Matthew Kubik; A sustainable and universally designed home renovation for a C-4 quadriplegic by Megan Lee (EDRA 40 Annual Conference 2009)	7
Memorial Sloan Kettering Cancer Centre, Brooklyn, USA 2008–2009	This and future projects by the Parsons New School for Design (Interior Design Program) involve students investigating the role of technology in diagnosis, treatment and recovery. The design ideas for the Centre included 'meditation spaces and individualized lighting motifs for each room, as well as the provision of patient art programming that would take advantage of Brooklyn's local cultural offerings' (Parson New School for Design n.d.). These ideas were considered by the Centre's architects as part of the final design. The goal for Parsons was to challenge the roles that interior designers can play in contemporary society by shaping the physical, emotional, and psychological experience for patients in ways that empower them	5, 3
Safe Spaces, Austin, Texas USA	Project H (builders and designers), working with the Helping Hand Home for foster children in Austin, sought to create 'engaging, comforting, and inspiring therapy solutions' for the children. Insertions aimed to create colourful and tactile environments in the quiet rooms that provide a therapeutic outcome. Textures, objects, light, sound, landscapes and materials research findings were integrated to achieve that end (Project H Design n.d.)	5, 2, 4

practitioners, who are working in this area, will contact the editors, to assist in building upon the initial survey. The examples listed also act as stimuli for others to envisage how they may be involved in the future and/or may assist in establishing awareness and possible collaborations. Each example is linked to one or more of the ten modes defined previously.

In summary, it was found that in relation to Social Justice or social responsibility and interior architecture:

- the majority of projects are conducted for research or as projects within educational institutions. In the latter case, students are engaged in in-depth studies of, or with, relevant communities to develop a design brief
- the community and user-based design principles are stressed

- because projects respond to community needs, they are not necessarily tied to particular disciplines and collaborations are essential for success
- most organizations describe members by the umbrella term, design
- project funding is normally from a variety of sources including community donations, government and volunteer organizations.

Although named interior architecture projects are difficult to identify, it is evident that interior architects have contributed to Social Justice projects. However, in many cases, the contribution is not linked to the discipline's core skills and knowledge. Therefore, it is pertinent that the potential roles will now be discussed by contextualizing interior architecture within the wider debate of social responsibility and social sustainability. A need for action becomes apparent.

Social Equity and Social Justice: Is there a Role for Interior Architecture in the Future?

In 1995 Victor Papanek, in his seminal work *The Green Imperative*, challenged designers by posing five questions that must be asked of their design. Two of these are relevant here: 'Can it [the design] make life easier for some group that has been marginalized by society? Will it help those who are poor, disenfranchised or suffering?' (Papanek 1995, p. 54). Nigel Whitely also notes that design 'provides a snapshot of society's condition' because it represents, or is symbolic of, what a society or community believes and the collective rules by which it operates, as well as the values that underpin these; and he argues that designers and consumers 'can no longer plead innocence or amorality' (Whitely 1993, p. 170); or, as John Thackara states, ignorance of the bigger picture is not an excuse (Kettles 2008, p. 50). If we consider what defines many disadvantaged or marginalized people, it is a secure shelter or home and/or the provision of basic services that are often missing. This may be in the context of their residence, but it may also be applied to other everyday activities that people engage in, such as work, healthcare, social interaction and education. Therefore, it would seem pertinent to inquire into the role and potential need for interior architects to be involved as key players in the improvement of living conditions for such people.

Importantly, the designers of interiors possess key knowledge pertaining to the relationship between people and the built environment. They potentially deal with the experience of people through the relationships fostered by design. They also understand the concept of occupation and how the characteristics of an environment—colour, material, light, space, volume, services, orientation, and the like—can impact on the quality of engagement. Even so, the profession of interior architecture is often mistakenly equated with interior decoration, and therefore, its value is perceived as limited to selecting surface treatments and furniture arrangements within existing or proposed environments. However, the role of interiors can extend from the smallest places to palatial buildings, into courtyards, passages, and a range of other environments where a sense of interiority is evident. The aspects of a design are understood by interior practitioners not simply in terms of their physicality, but also as an extension of daily rituals, whether for an individual, family, community or society. As a result, interior experts understand how design influences identity, belonging and levels of discrimination.

In 2005 I outlined how the built environment was and is discriminatory. Discrimination is defined as: 'a point of distinction, and in relationship to people in environments … [it] also extend[s] to notions of inclusion, exclusion, appropriateness, and inappropriateness, to belonging, and to a sense of rejection' (Smith 2005, p. 64). The way the environment achieves this includes:

> … an ability to subsume a person's identity, influence identity construction, challenge codes relating to a person's appearance, influence the definition of roles and the associated expectations, and influence the level of inclusion or engagement through the way the interpreter and interpreted are positioned. (Smith 2005, p. 72)

And I concluded that '… we can no longer look at the physical environments, social environments, and the individual experience as separate from one another if we wish to understand everyday life' (Smith 2005, p. 72).

The research developed Leslie Weisman's observation in the seminal work *Discrimination by Design*, which clearly demonstrated how the design of interiors, such as birthing wards in hospitals, relate to issues of power and social order (Weisman 1994, p. 23). Kristine Jerome's study of women's refuges also demonstrates how social order can be maintained or challenged through layout, and therefore, the associated rituals (Jerome 2002). These examples reinforce the importance of addressing the needs of those who are marginalized within cities and developed nations as well as those in rural and remote regions, and within developing nations. They show that both the potential and the need to address occupation through design for interior architects are strong.

However, design decisions can further disadvantage the already disadvantaged. For example, Whitely, in 1993, observed that shopping mall locations can 'disadvantage low-income women, the majority of whom are working-class and/or black, because these women lack access to private transport' (Whitely 1993, p. 159).

The Need

It is understandable that in areas of high level disaster and trauma, those who are best able to address the construction of services and infrastructure are needed at the forefront of action. People with skills pertaining to engineering and building, along with those dealing with sanitation and medicine, are often the first to be brought on board.

However, when dealing with people whose circumstances construct their mid- to long- term realities, the skills and knowledge of an interior expert have the potential to be of great value. Attributes to consider in planning habitats in these situations include:

- awareness of how the environment influences orientation, navigation and travel or mobility
- understanding of how the environment influences mood and atmosphere
- understanding function, activity and the associated physical and spatial requirements
- understanding of how the environment influences image and identity and sense of self and/or community
- awareness of environment as an enabler of self expression
- awareness of environment as a facilitator of levels and degree of interaction
- understanding of how the environment engenders or inhibits a sense of privacy
- understanding of the interdependence of the person and the environment as a whole and its constituent parts
- understanding of the interplay of meaning and purpose for individuals and collectives
- knowledge of rituals and their physical, social and psychological contexts
- sensitivity to people's needs and an ability to communicate.

In health-related environments, interior professionals can extend such knowledge to create therapeutic environments.

While undertaking a post-occupancy study of a residential block for young males with brain injury in Queensland, Australia, it was evident that the environment was important for many of the above reasons. For example, ventilation and orientation influence the temperature as well as comfort levels, thereby raising the potential for anti-social behaviours to develop in uncomfortable settings; street location influences access to shops, services and transport, thereby influencing a sense of independence; having individual mailboxes, ensuites and private rooms create a sense of the person's identity; while a choice of areas to interact with other residents or friends enables socializing (Smith and Adkins 2006).

Creating a Framework to Facilitate Engagement

Learning from Others

Miles Roston, relating the story of Siri—a young woman from Thailand who helped a woman give birth in the back of a pick-up truck and as a result, studied biology, became a teacher, and set up a teachers' network to assist children—shares Siri's principles, which evolved through her work and her aim to understand the needs of people who are 'in times of want' (Roston 2010). Her principles are:

- To look: to see the problem or dilemma and allow oneself to experience it.
- To listen: to hear how the person or community thinks the problem can best be approached, rather than imposing one's own solution.
- To use what you know: each of us has a skill that can help.

- To have patience, because short-term change changes back.
- To give encouragement: even more crucial than throwing money at a problem, though money is necessary too, is to inspire hope and vision, and to encourage people to dream (Roston 2010, pp. 25–26).

These principles emphasize the value of giving whatever we have to share through a process whereby we truly listen, and therefore come to understand where our gifts can be best used and why. These principles are obviously relevant for the broader discussion of Community Engagement found in Chapters 2 and 3.

The danger of seeing everyday circumstances as problems can influence designers' approaches and attitudes. As Christopher Day says of architecture:

> … design proposals were referred to as 'solution'. People can be offended if you refer to their home as a problem, or a solution—to them it is a living being, rich in multiple functions both spiritual and practical, mostly inseparable. The simplistic categorical nature of diagram thinking and problem solving tends to push the qualitative aspects into second place at least … There is also a resultant disconnection from the subtleties of site and context during the design process (Day 1999, p. 9).

The outcome of a design when occupied can also impact on the occupants. For example, sick-building syndrome was conceptualized as it became evident that certain materials and services triggered health problems such as asthma, once the buildings were occupied. Post-occupancy evaluations strove to identify the links between environmental indoor quality and user satisfaction and productivity. The work of W. Preisner, in the 1980s, spawned an industry that is still evolving today (Preisner 1981; Preisner and Vischer 2005). As Day points out, the nature of the interior influences the ways in which light affects the quality of the space occupied and how it is experienced; for instance, conditions such as inadequate lighting have been linked to depression (Day 1999, p. 121).

Understanding and being responsible for the potential outcomes of the profession's and our personal practice will enable interior architecture to play a significant role alone and as part of collaborative projects.

The Shifting Agenda

There is a recent shift toward consideration of other agendas beyond design for design's sake: design for profit, design that supports consumption, and/or design that reinforces inequities. As Whitely pointed out nearly two decades ago:

> These values imply a commitment not only to the current fashionable notions of 'integrity' and 'honesty', but also to societal notions of 'equality' and 'justice'—terms which seemed archaic in the 1980s but which are now turning into political agendas. The balance between private affluence and social responsibility does seem to be beginning to shift more towards the latter. (Whitely 1993, p. 161)

However, at that time Whitely also posited that the design profession or design press did not engage a discussion of such values explicitly, and challenged the assumption that good design needs to be defined in terms of commercial success (Whitely 1993).

The same debate is relevant today, but in many professions there is a greater awareness and a desire to contribute socially and responsibly. More recently the concept of Corporate Social Responsibility has evolved from the classic model to a socially-responsible model. It has transformed from 'the management's only legitimate goal is to maximise profit' or 'corporate expenditure on social causes is a violation of management's responsibility to shareholders, at least to the extent that these expenditures do not lead to higher shareholder wealth', to become a model where 'a firm's management is responsible for maintaining an equitable and working balance among the claims of the various directly interested groups such as stockholders, employees, customers and the public at large' (Othman 2009, p. 44).

A variety of recent projects recognize the importance and value of locale and the human aspects. For example the movement Slow Design aims to create objects that 'nourish and inspire' and that 'carry with them a story of how they were conceived, made and inspired' (Kettles 2008, p. 48). Allistair Fuad-Luke states that Slow Design fosters the 'emerging slow movement' and describes the six principles of SlowLab—reveal, expand, reflect, engage, participate and evolve; and as a result there is a transformation in how designers work (Fuad-Luke 2009, pp. 157–158). The Design for the other 90 % project, curated by Cynthia E. Smith in 2007, drew people from across the globe to generate ideas that addressed the needs of some of the poorest people in the world. Smith quotes that at that time 2.8 billion people lived on less than two dollars a day (Wagner 2007, p. 196).

The exhibit's emphasis, however, was largely on product design rather than spatial design, or more specifically, interiors. Her book includes issues which are interior-relevant, such as Informal Community Solar Kitchens and the Katrina Furniture Project (Smith 2007). In association, publications such as Design Like You Give a Damn, edited by Diana Murphy for Architecture for Humanity (Murphy 2006), brings into focus the work being done by architects and, in association, spatial design. As Joe Holyoak observes, such works challenge us to see how design can 'assist societies to cope with poverty, disease and disaster' (Holyoak 2006, p. 224). There is a consistent belief by people and organizations working in this domain that local communities must be part of the process and the outcomes. M. Jann and S. Platt describe architectural projects, including the work of Fathy, which enabled locals to make materials and build while advocating the 'trinity' of owner, architect and craftsman (Jann and Platt 2009). Jones, similarly, advocates that locals should identify the project needs and set the parameters, including the brief (Jann and Platt 2009). These practitioners also stress the importance of partnerships in increasing relevance and impact. For example, the development of a residential campus for former street children and women in distress, Amatitlan, Guatemala, involved an interdisciplinary, multicultural partnership including:

> Stanford University architectural design class, Studio IMPACT (International Management Projects in Architectural Construction Technology), and an MBA social entrepreneurship workshop led by Jann at the Ecole Nationale des Ponts et Chausses in Paris. This international collaboration integrated design sensitivity with professional expertise, local know-

> how, research, and 'social ethics' to respond to acute poverty and need. The program for the project included a clinic, small community/administration centre, workshop, and school, with the residential model based on simulated nuclear family units accommodating 6–8 children each.
> ...
> Another, the 'Center for the Restoration of the Multitudes (CEREM)', founded by Estela Solis, provides shelter, drug treatment, and therapy for former gang members and prostitutes (Jann and Platt 2009, pp. 86, 87).

Lessons and Limitations

Although such projects provide suggestions regarding how to contribute meaningfully to the social design agenda, there are also limitations raised that the interior professions need to learn from. For example, a survey by A. A. E. Othman identified four obstacles to South African architectural companies adopting a Corporate Social Responsibility (CSR) approach (Othman 2009). These are:

- lack of integrating CSR in the internal governance of design firms
- limited government incentives for supporting CSR and the negative perception of time
- cost
- energy constraints when CSR is adopted.

Suggested strategies Othman identified to counteract these include:

- raising awareness within the profession and becoming more proactive rather than responsive
- adopting social responsibility into the firm's mode of operating or governance
- monitoring effectiveness of actions adopted
- appointing an ethics officer to manage implementation and liaison with community
- publishing, exhibiting and publicizing successes that have solid evidence (Othman 2009, p. 43).

In terms of the process when undertaking such projects, Jann and Platt also provide insights from their overview of Philanthropic Architecture that are relevant to other disciplines such as interior architecture. Principles include:

- effective involvement of design NGOs and Western designers/architects with local community members in true collaborations
- advanced technical acumen, as poverty leads to difficult sites and construction needs
- small design teams (preferably six to eight) with roles clearly articulated early between the locals who are in the client group, design team, construction team or involved in project management
- when arriving at an impasse with fundraising, it is useful to put finished design projects on the back burner to "percolate" while turning to other projects

- a clear, realistic timeline and technical competency are both critical for the project
- in order to deal with inevitable conflicts, a need to: 'stay focused on a common goal, rise above personality conflicts and egocentrism, value each other's experience and skill sets, respect differences in customs and backgrounds, communicate with each other, cultivate team spirit, and be willing to work through issues'
- an awareness of the limitations and ensure project scoping matches the available resources at the project location (Jann and Platt 2009, p. 90).

Where to From Here?

Consider the working groups or steering committees associated with some of the most pressing environmental projects—most recently, tsunamis, volcanic activity, earthquakes, floods and refugee movements. It is most likely that we will expect these bodies to include policy makers, urban designers, planners, architects, social scientists, NGO representatives, medicos and the like. Therefore, is the absence of expert interior architects a lost opportunity?

This chapter has provided insights into how interior professionals can be involved in socially responsible design. It highlights the core attributes of the profession and challenges those who see the current contribution of the profession to be limited by an understanding of the interior world as just applied finishes. It also raises issues that can help inform the development of our involvement by introducing the work of other professionals working in this field.

In addition, the agenda set by the IFI is a stimulus as well as challenge for the professional bodies globally as well as for students, academics and practitioners. The shifting local attitudes discussed also raise new possibilities.

The practitioners selected to share their experiences in Chapters 5 to 8 serve both to demonstrate what is occurring at this time and to introduce a number of issues critical to the contemporary debate in this area.

Acknowledgement I would like to acknowledge the valuable input of research assistant, Rachael Carter, who was responsible for the collection and collation of the data during the project.

References

American Society of Interior Designers. (n.d.). http://asidfoundation.We-Care.com/Start. Accessed Sept 2010.

Architecture for Humanity. (March 2010). Structures for inclusion 10 conference. http://architectureforhumanity.org/events/2010-03-27-structures-for-inclusion-10-conference.

Art and Design for Social Justice Symposium. (n.d.). http://socialjusticesymposium.fsu.edu/. Accessed Sept 2010.

Australian Red Cross. (n.d.). http://www.redcross.org.au/default.asp. Accessed Sept 2010.

Butterfly Works. (n.d.). Creative volunteers. http://www.butterflyworks.org/content/16453/creative_volunteers. Accessed Sept 2010.

Changing Minds. (2010). http://changingminds.org/explanations/needs/maslow.htm. Accessed Sept 2010.
CODA-STUDIO. (n.d.). http://www.coda-studio.com/.
Cooper-Hewitt National Design Museum. (n.d.). Design for the other 90%. http://other90.cooperhewitt.org/. Accessed 11 March 2011.
Day, C. (1999). *Places of the soul: Architecture and environmental design as a healing art*. London: Thorsons.
Department of Families, Housing, Community Services and Indigenous Affairs (FaHSCIA). (2009). Recent international and national approaches to homelessness. Appendix one: Statistics on homelessness by European country. http://www.fahcsia.gov.au/sa/housing/pubs/homelessness/saap_er_publications/recent_international_national/Pages/appendix1.aspx.
Design Corps. (2009). Community design and planning. http://www.designcorps.org/projects/community-design-planning. Accessed Sept 2010.
Designers Institute of New Zealand. (2003). Code of Ethics. http://www.dinz.org.nz/content/10529/code_of_ethics.pdf. Accessed Sept 2013.
Drexel University. (2008). Design charrette. http://www.drexel.edu/westphal/charrette/. Accessed Sept 2010.
Environmental Design Research Association (EDRA). (2009). The ethical implications of sustainability in interior design. EDRA 40th Annual Conference. http://www.edra.org/content/ethical-implications-sustainability-interior-desig. Accessed Sept 2010.
European Charter of Interior Architecture Training. (2007). http://www.ecia.net/content/european-charter-interior-architecture-training-2007-now-available-6-languages. Accessed 22 Nov 2010.
European Council of Interior Architects (ECIA). (n.d.). http://www.ecia.net/what_we_do/code_of_conduct/english. Accessed Sept 2010.
Fuad-Luke, A. (2009). *Design activism: Beautiful strangeness for a sustainable world* (pp. 157–158). London: Earthscan.
Holyoak, J. (2006). [Review of the book *Design like you give a damn*, by C. Sinclair and K.Stohr]. *The Architects' Journal, 45*(3), 224.
Homelessness Australia. (2010). http://www.homelessnessaustralia.org.au/UserFiles/File/Homelessness%20in%20Australia%202010%20lr(1).pdf. Accessed Sept 2010.
Human Rights Council, United Nations General Assembly. (2009). Report of the special Rapporteur on adequate housing as a component of the right to an adequate standard of living, and on the right to non-discrimination in this context, by Miloon Kothari.
International Federation of Interior Designers (IFI). (n.d.). http://www.ifiworld.org. Accessed Sept 2010.
International Interior Design Association. (n.d.). http://www.iida.org/content.cfm/iida-policy-statements. Accessed Sept 2010.
Interior Design. (2010). Parsons provides inspiration for Memorial Sloan-Kettering Cancer Center Alternate Design Models. http://www.interiordesign.net/article/478649-Parsons_Provides_Inspiration_for_Memorial_Sloan_Kettering_Cancer_Center_Alternate_Design_Models.php. Accessed 6 Mar 2010.
Interior Design Educators Council (IDEC). (2010a). Interior design educators council nomination for the IDEC community service award 2010. http://www.idec.org/news/documents/CallforNominationIDECCommunityServiceAward.pdf. Accessed Sept 2010.
Interior Design Educators Council (IDEC). (2010b). Make a difference. http://www.idec.org/students/MakeaDifference.php. Accessed Sept 2010.
Interior Design Educators Council (IDEC). (2010c). Newsletter. http://www.idec.org/news/documents/IDEC-Newsletter_2010SeptemberIssue.pdf. Accessed Sept 2010.
Jann, M., & Platt, S. (2009). Philanthropic architecture: Nongovernmental development projects in Latin America. *Journal of Architectural Education, 62*(4), 82–91.
Jerome, K. P. (2002). Constructing independent women through space: A case of the women's refuge. *IDEA Journal 2002,* 52–70.
Kettles, N. (2008). Designing for destruction. *The Ecologist, 38*(6), 48–50.
Kids' Guernica. (n.d.). http://www.kids-guernica.org/. Accessed Sept 2010.

Kothari, M. (2007). Addendum: Mission to Canada. http://www.equalityrights.org/cera//wp-content/uploads/2010/03/Sp-Rapp-Housing-Report-Canada.pdf. Accessed Sept 2010.

Martin, C. S., & Guerin, D. (2005). The interior design profession's body of knowledge. http://www.careersininteriordesign.com/idbok.pdf. Accessed Sept 2010.

Murphy, D. (Ed.). (2006). *Design like you give a damn: Architectural responses to humanitarian crises*. New York: Distributed Art Publishers, Inc.

NCU Center for Service Learning. (n.d.). http://www.wcu.edu/9815.asp. Accessed Sept 2010.

Ngari School Construction project. (n.d.). http://www.architecturemedia.com/aa/aaissue.php?issueid=201009&article=19&typeon=2. Accessed Sept 2010.

Othman, E. (2009). Corporate social responsibility of architectural design firms toward sustainable built environment in South Africa. *Architectural Engineering and Design Management, 5*(1–2), 36–45.

Papanek, V. (1995). *The green imperative*. London: Thames and Hudson.

Parsons The New School for Design. (n.d.). Memorial sloan kettering cancer center. http://www.newschool.edu/parsons/memorial-sloan-kettering-cancer-center/. Accessed Sept 2010.

Preisner, W. (1981). *Post occupancy evaluation*. New York: Van Nostrand.

Preisner, W., & Vischer, J. (2005). *Assessing building performance*. Oxford: Elsevier Butterworth-Heinemann.

Project H Design. (n.d.). About safe spaces. http://projecthdesign.org/projects/safespaces.html. Accessed Sept 2010.

Roston, M. (2010). *Making a world of difference; inspiring stories of the world's unsung heroes*. Wollombi: Exisle Publishing.

Rowan, J. (1998). Maslow amended. *Journal of Humanistic Psychology, 38*(1), 81–92.

Smith, D. (2005). Environmental distinctions: The discriminating dining environment. *Cahiers du CICLaS, 6,* 53–73.

Smith, C. E. (2007). *Design for the other 90%*. New York: Cooper Hewitt, National Design Museum. Smithsonian Institution.

Smith, D. (2009). Spatial design as a facilitator for people with less visible impairments. *Australian Medical Journal, 1*(13), 220–227.

Smith, D., & Adkins, B. (2006). *Post occupancy evaluation report: Street residences*. For the Brisbane Housing Company (unpublished).

The Salvation Army. (2010). Submission prepared by the Salvation Army Southern Territory in response to the 'Regulation and growth of the not-for-profit housing sector' discussion paper. http://www.sarmy.org.au/Global/SArmy/Resources/social/salvation_army_nfp_housing_submissionl_june_4_2010.pdf. Accessed Sept 2010.

The South African Institute of Design Professionals (iid). (2010). Letter from the IID President. Inside the lid this month, 28, 8. http://www.iidprofessions.com/documents/iidinsightmarch2010.pdf. Accessed Sept 2010.

The United Nations Educational, Scientific and Cultural Organization (UNESCO). (n.d.). http://portal.unesco.org/culture/en/ev.php-URL_ID=35418&URL_DO=DO_TOPIC&URL_SECTION=201.html. Accessed Sept 2010.

Torres, Personal Communication.(2010). Powerpoint presentation, 2010 for Colmeia Community Project (Associação de Amigos do Projeto Colmeia), Project Outline-September 2010.

University of South Australia, Centre for Settlement Studies. (n.d.). http://w3.unisa.edu.au/css/. Accessed March 2011.

Urban Studio. (2007). http://152.13.13.150/iar/us/.

Wagner, G. (2007). Designing for the "Other 90 Percent". *The Futurist, 41*(3), 196.

Waxman, L. (2007). Designing the third place: Enhancing social opportunities for older adults. Proceedings of the 2nd Annual Arts and Design for Social Justice Symposium. Florida State University, Tallahasee, FL. Accessed Sept 2010.

Weisman, L. K. (1994). *Discrimination by design: A feminist critique of the man-made environment*. Urbana: University of Illinois Press.

Whitely, N. (1993). *Design for society*. London: Reaktion Books.

Wright, N. (2010). DIA design: Action presentation, Central TAFE, Perth.

Chapter 5
The Importance of Interior Design Professionals' Engagement in Policy Development to Effect Change for Underprivileged Individuals

Jill Pable and Lisa Waxman

Abstract The authors propose that designers can contribute toward design (and associated policies) for disadvantaged individuals in meaningful ways and discuss the potential that environmental psychology holds for the development of low-income housing. Increasingly, the intricate situations of our culture require systemic approaches to addressing problems in ways that complement the core skills and knowledge that designers possess. Designers have an ability to think broadly about problems and to create innovative approaches to contemporary issues. A Florida Housing Finance Corporation (FHFC) project is discussed as an holistic model of support which integrates inclusive methods of solution-seeking. In an effort to expand their guidelines for low-income supportive housing to include more human-focused design considerations, the FHFC approached the university to seek out the authors, who had the knowledge to help them with this endeavour.

As expected, we found that issues such as privacy, safety and security, and territoriality were very important to the residents. We also found that residents wanted to be proud of the place they where living and that they were pleased when they developed a sense of attachment to that place and to the other residents and staff who resided or worked there. In addition, it is evident that the process involved, not so much the need for more accessibility standards etcetera (although these are essential), as a way to holistically address the positive, healing nature of the built environment for comprehensive resident satisfaction—mentally, socially and physically. In order to facilitate this, environmental psychology offers a wealth of strategies applicable to supportive housing environments. The study has shown us that designers have the ability to offer insightful advice to the very real problems of low-income housing, and that knowledge from alternative disciplines can assist.

Keywords Affordable housing · Systematic approaches · Human-focused design · Environmental psychology · Interior design

J. Pable (✉) · L. Waxman
Department of Interior Design, Florida State University, Tallahassee, FL, USA
e-mail: jpable@fsu.edu

L. Waxman
e-mail: lwaxman@fsu.edu

D. Smith et al. (eds.), *Perspectives on Social Sustainability and Interior Architecture*, DOI 10.1007/978-981-4585-39-2_5, © Springer Science+Business Media Singapore 2014

Introduction

United States President Jimmy Carter remarked that 'the measure of a society is found in how they treat their weakest and most helpless citizens' (Carter 2010). Undoubtedly a critical aspect of that treatment is housing, which represents one of the most elemental needs of life. In the United States, residences for the economically disadvantaged are often built by private developers and through government programs that provide loans and tax incentives for housing developments. This essay discusses the potential that environmental psychology holds for the betterment of these low-income housing developments, and by association, how designers can contribute toward policies for disadvantaged individuals in meaningful ways.

We speak of this topic from the basis of a project in which we are currently engaged in the southeast part of the United States with a government entity called the Florida Housing Finance Corporation (FHFC). This body is charged with awarding loans to developers who qualify for them through their adherence to various codes and standards requirements mandated by FHFC and others. These range from general building codes that describe minimum square footages and plumbing requirements, to sustainability guidelines that mandate the presence of ceiling fans and energy-saving appliances. However, FHFC is considering altering their approach to low-income housing from one that was strictly 'sticks and bricks' to a more holistic framework that better reflects the needs and identities of low-income residents such as the frail elderly, veterans, and the recently homeless. FHFC suspects that this holistic approach to resident wellbeing may logically extend to the design of the residential built environment. While FHFC has already started including requirements that tap a diversity of issues including accessibility, sustainability and visitability—the means for the physically disabled to visit others in their homes—FHFC staff have sought recommendations that scope beyond these existing standards systems into other areas that provide meaningful support for the human condition.

FHFC's trend toward a more holistic model of support mirrors the broader movement toward inclusive methods of solution-seeking, perhaps brought on by the acknowledgement of the complexity of the problems that face our culture. Sim Van der Ryn, for example, suggests Western culture is moving from the 'mental age' characterized by fragmented, solitary thinking into the 'integrated age', known for collaborative reasoning, personalization, and close user–client connection (Van der Ryn 2007). Similarly, others see that the increasingly intricate problems of our culture imply, and even demand, that 'systemic' approaches to problems be undertaken (Levy 1990). This is pertinent to our discussion here, as some suggest that designers possess the largely unique ability to think broadly about problems; this permits them to serve as intercessors of sorts between current ways of thinking and the use of new, necessarily comprehensive approaches to achieve solutions (Diani 1998). Such systematic thinking has already enhanced approaches to more effective solutions to problems that affect the disadvantaged, including emergency shelter (Tuvie 2010)· access to clean water (Vestergaard 2013) and food preservation (Rolex Awards for Enterprise 2000).

Taking cues from others' successes along these holistic lines, we intend to address resident needs by tapping established research from environmental psychology and applying these concepts to the requirements of homeless, veteran and frail elderly residents. In so doing, we envision a series of recommendations that address general wellbeing and embrace previously neglected mental needs such as place-making, self-identity, sense of community and privacy. Such action may represent first steps towards extending the current concept of liveability—which is currently applied to neighbourhood-scope strategies such as sense of community, sense of place, and security—to the interior human-environment relationship within a residence and associated community spaces.

In an effort to expand their guidelines for low-income supportive housing to include more human-focused design considerations, the FHFC decided to look beyond their own organization for assistance. Located in the same town as a Florida State University, the FHFC sought out professors who had the knowledge to help them with this endeavour. Once contacted, we were intrigued at the realization that new human-focused guidelines could impact thousands of low-income supportive housing units throughout the state. After some discussion, we became aware that although we both had some background in environmental psychology as it relates to interior design, we needed to get out into the field and see the supportive housing units, and talk with residents, support agency representatives for the homeless, veterans and the elderly, and developers. We reasoned that first-person findings in the field would logically help us avoid the pitfalls that could occur were we to generate our ideas solely from the academy.

The project began in autumn 2009 with planning, followed by site visits in the spring and summer of 2010 to various low-income supportive housing developments throughout the state. We conducted telephone and face-to-face interviews with developers, support agency representatives and residents to gather their experiences of building low-income housing, servicing the needs of residents in low-income housing, or their experiences as residents living in such housing. We also took photographs to visually document existing conditions.

Once the visits, interviews and visual documentation were complete, we used qualitative research analysis to evaluate our field notes and interview data, and coded and analysed the findings. A number of overarching themes emerged which were related to residents' emotional and physical needs—factors, we found, that were often interwoven. We realized how important it was to understand the state of mind of the resident from the time they first arrive in supportive housing—often fragile and scared—until they move on to another housing situation, hopefully, stronger and better equipped to handle the next phase of life.

As expected, we found that issues such as privacy, safety and security, and territoriality were very important to the residents. However, several other interesting themes emerged, including:

- a strong desire for continuity in housing (that is, knowing how long you can stay and where you'll be going)
- a desire for housing that looks like 'everyone else's' in the neighbourhood

- needs for services that are offered on-site (for example, health fairs, support group meetings)
- need for support staff on-site with adequate space to complete their work (for example, caseworkers and counsellors)
- need for autonomy (for example, to make a meal for oneself)
- need for places to be and feel productive (for example, computer labs and study spaces)
- need for group gathering spaces to help create feelings of community (for example, to celebrate holidays and birthday parties)
- the desire for spaces in which unforced encounters and friendship formation can take place (that is, creating a support network).

We also found that residents wanted to be proud of the place they were living and that they were pleased when they developed a sense of attachment to that place and to the other residents and staff who resided or worked there.

It is important to note that particularly useful information came from agency representatives who provide services to these populations. We found them to be in touch with the needs of the residents as well as being a rich source of many realistic suggestions for the design guidelines. During interviews, residents often reported that their existing housing was fine—even perfect—although on-site observation suggested otherwise. We eventually realized, like other researchers who work with the disadvantaged, that residents were unwilling to critique their environments, either because they were afraid to complain for fear of negative consequences and a potential loss of housing—even with assurances of their anonymity—or, alternatively, because they suppressed the need for experiences they perceived they could never have (Banyard and Graham-Bermann 1998; Douglas 1976). Nonetheless, helpful advice from many residents urged us to retain and embrace these comments, which in turn enhanced the study's cumulative guidelines.

What we found was not so much the need for more accessibility standards etcetera (although these are essential), but more a way to holistically address the positive, healing nature of the built environment for comprehensive resident satisfaction—mentally, socially and physically. Environmental psychology offers a wealth of tested, considered strategy that can be applied to supportive housing environments.

The outcome of this study is still in progress, but is emerging as a series of practical, supported design recommendations, ranging from adequate storage and counters near bathroom vanities, to large community rooms and offices for support staff. To make it easier for developers to understand the recommendations, we are creating a booklet of visual guidelines explaining the problem and offering a solution that includes graphic examples. (Figs. 5.1 and 5.2)

The study has shown us that not only do designers have the ability to offer insightful advice to the very real problems of low-income housing, but that these ideas can bring into play proven but previously untapped content from new sources like environmental psychology, that can address residents' needs comprehensively. We are hopeful that our summative report to FHFC will be well received and aspects of its recommendations integrated into developer loan guidelines.

Bathroom Mirror Lighting and Color

Context of the Problem

When we check and adjust our appearance at the mirror, lighting and color play a key role in what we see, how we appear, and to a certain extent, how we feel about ourselves. That is, incorrectly placed or color temperature lighting can present unflattering mirror views, making one appear more tired or weathered than daylight conditions provide, which is the general yardstick against which we assess the reality of color and texture. Similarly, wall color near mirrors can deeply affect reflected skin coloration, presenting a deceptive and oftentimes unflattering self-perception.

Why this Matters

Self-image and positive self-esteem are key to full participation in society. Further, researchers suggest that positive self-image is connected to success in acquiring and maintaining employment (Burn, 1992). People are very attuned to the messages sent by various skin tone colors, with pinkish skin denoting health and vitality, and skin in greenish or bluish tones associated with illness or weariness. The color of interior lighting can make the healthiest individuals' skin tones look unnatural, and by extension, unhealthy. Unnatural lighting colors can also make matching clothing, such as socks, difficult.

The placement of lighting in relation to a mirror is also important. This is because light fixtures often cast shadows, and if those shadows are under the nose or highlighting bags under the eyes, this affects self-perception. Lighting that is placed high at the mirror and is bright and direct is apt to create such shadows and is also difficult to relamp, given its ceiling location.

Another hazard is lighting fixtures that are too bright and glary. These cause strain and irritation to the eyes, as the eyes try to continually adjust between bright areas and dark areas. Problems such as these with interior lighting often can have the affect of subtly altering mood or outlook as one prepares for the day.

Persons who require housing assistance may be particularly fragile with regard to self-esteem. Therefore, an environment that provides as much support as possible for one's self-assessment of physical appearance is a practical idea.

Solution

Be careful with wall coloration, lighting color, and lighting fixture placement in areas connected to mirrors, particularly in bathrooms where preparation for the day often occurs. The best lighting for these situations is placed on either side of the mirror and is a style that uses a fluorescent or LED lamp (for sustainability and longevity) and has a diffusing cover that prevents glare. Fixtures that permit easy relamping are also best, as are those with acrylic or other plastic translucent diffusers that increase durability. Lamps that are 3500 degrees Kelvin (color temperature) offer the most realistic light color (similar to daylight)[1].

In small bathrooms, lighting fixtures can be integrated wtih the mirror, such as that shown below[2].

Avoid wall colors near mirrors that are in the blue or green hues, as this negatively affects the appearance of skin color.

[1]Visa Lighting offers an example of a small profile acrylic fixture at http://www.visalighting.com/models/CB6504/

[2]Pegasus Lighting offers a mirror that integrates fluorescent lighting for small spaces. See http://www.pegasuslighting.com/vanity-mirror-lighted-square-fluorescent-110w.html

Fig. 5.1 An example recommendation for bathroom lighting proposed for FHFC suitable for the study's homeless, veteran and frail elderly users. (Image by Jill Pable)

Solution

Transitional Housing for the Recently Homeless
In-Room Sinks, Vanities & Mirrors

Provide a bed/bathroom design arrangement for transitional housing for the homeless that separates the vanity/sink/mirror from the room containing the toilet and bathtub.

The sink is used more than the other bathroom fixtures not only for water, but also to check one's appearance, applying makeup and similar time-consuming activities. By placing the vanity and sink with mirror within the bedroom, the family has access to it at their leisure and can store their hygiene products there. The vanity /mirror area should have quality lighting with a minimum color rendering index (CRI) of 80.[1]

The sink should be in a vanity cabinet (not a pedestal sink) that offers a horizontal counter and with storage underneath. A wall-recessed medicine cabinet also provides needed storage above the vanity. Accessible rooms will require knee space underneath the sink for use.

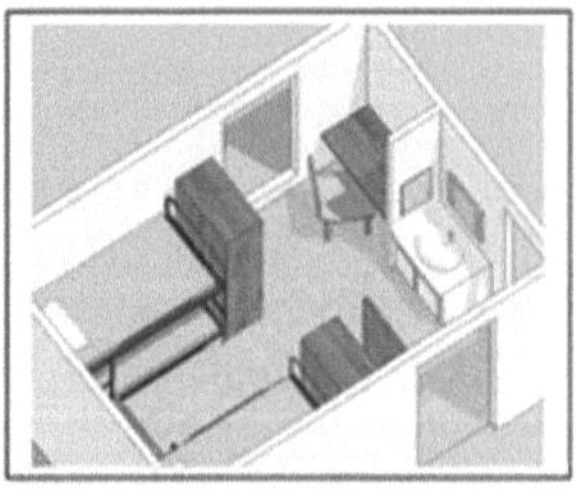

A transitional shelter bedroom showing an in-room sink, vanity and mirror with lighting.

[1]http://www.gelighting.com/na/business_lighting/education_resources/learn_about_light/color_rendering.htm

Context of the Problem

Transitional housing developments for the recently homeless often have an arrangement of bedrooms for 2 or more individuals that share a bathroom with another similar room. Thus, often up to eight people share a single bathroom with a sink, toilet and shower/tub. While this is economical, it also creates issues for residents:

- Residents are forced to store their hygiene products in their rooms for real or perceived security reasons. This adds to the clutter of these often small rooms.
- Use of the sink can unnecessarily tie up access to the toilet and tub/shower

Why this Matters

Those that are recently homeless have particularly strong connections to those few personal belongings they still have, and this includes consumable products such as bathroom supplies. Thus, it helpful to provide residents the means to secure their possessions simply and easily. It is also important to provide these persons with the means to personally control their environments which may in turn nurture empowerment-- at a time when self-esteem is often low.

The control of a mundane part of the private environment such as a sink/vanity/mirror can ease the process of multiple persons in a single room getting prepared for their day, many of which may be getting ready for school or work at the same time. It can also provide residents with the dignity of enhanced access to their sink/vanity for their family without sharing and can provide extra secured and easily accessible storage within the room.

Recently homeless: transitional housing Storage and Access

Fig. 5.2 An example recommendation for space planning proposed for FHFC suitable for the study's homeless users in transitional-style housing developments. (Image by Jill Pable)

References

Banyard, V., & Graham-Bermann, S. (1998). Surviving poverty: Stress and coping in the lives of housed and homeless mothers. *American Journal of Orthopsychiatry, 68*(3), 479–489.

Carter, J. (2010). Wisdom on-demand. http://www.iwise.com/8nXU5. Accessed 27 Oct 2010.

Diani, M. (1998). Immateriality takes command. *Design Issues, 4*(1/2), 6–11.

Douglas, J. (1976). *Investigative social research.* Beverly Hills: Sage.

Levy, R. (1990). Design education: Time to reflect. *Design Issues, 7*(1), 42–52.

Rolex Awards for Enterprise. (2000). The desert refrigerator. http://www.rolexawards.com/explore/themes/elegant_solutions#&!/elegant_solutions/the_desert_refrigerator. Accessed 13 Nov 2013.

Tuvie design of the future. (2010). Uber shelter: An emergency shelter in disastrous events. http://www.tuvie.com/uber-shelter-an-emergency-shelter-in-disastrous-events/#more-763. Accessed 27 Oct 2010.

Van der Ryn, S. (2007, May 31). *Surpassability: The integral age and resilient society.* Keynote speech at the Environmental Design and Research Association Conference. Sacramento, CA.

Vestergaard. (2013). Lifestraw carbon for water program launched in Kenya. http://www.vestergaard.com/news/press-releases/item/lifestraw-carbon-for-water-program-launched-in-kenya-2. Accessed 13 Nov 2013.

Chapter 6
Interior Design by Expediency

Rodney Harber and Angela Buckland

Abstract This is an account of South African migratory workers who still live an isolated existence in huge single sex hostels, far from their families. It describes the undignified conditions they are forced to confront, and their ongoing battle to express their identity on and around their stark bed spaces. The evidence from South African hostels illustrates how residents have responded with expediency to facilitate bed-based livelihoods, and proclaims the importance of expressing their individuality and personal dignity. These facilities have also proven to be problematic since they are open to manipulation by political parties. The most compelling argument against them is the spread of HIV/AIDS from 'town wives' via the men on their return visits to their families in the hinterland. However, high-density inner city accommodation has merit. Hence, if similar institutions are to remain, the lessons outlined in this essay are only a starting point. The impact of design (as well as no design) has far reaching consequences. In contrast to most interior design commissions, where we operate by creating images to coddle the relative wealthy, this is minimalist, functional, intensely personal and expedient strategy to create meaningful environments for the occupants.

Keywords Design · Interior design · Hostels · Identity · Migratory workers

Introduction

This is an account of South African migratory workers who still live an isolated existence in huge single sex hostels, far from their families. It describes the undignified conditions they are forced to confront, and their ongoing battle to express their

R. Harber (✉)
Harber & Associates, Architects, Urban & Rural Planners, Durban, South Africa
e-mail: rodneyharber@mweb.co.za

R. Harber
Honorary Research Fellow, University of KwaZulu-Natal, Durban, South Africa

A. Buckland
Angela Buckland Photographer, Durban, South Africa
e-mail: angelabuckland@yebo.co.za

D. Smith et al. (eds.), *Perspectives on Social Sustainability and Interior Architecture*,
DOI 10.1007/978-981-4585-39-2_6, © Springer Science+Business Media Singapore 2014

identity on and around their stark bed spaces. In contrast to most interior design commissions, where we operate by creating images to coddle relative wealth, this is minimalist, functional, intensely personal and expedient.

South Africa has had an unforgivable social history of excluding indigenous African workers from living permanently in towns and cities. Since colonial times, there were manipulative laws to force men to provide manual labour, to earn wages to pay Poll Taxes, but also excluding them from permanent residence by introducing a system of Pass Laws. These restricted town access to registered workers. I have enduring childhood memories from Pietermaritzburg of the curfew bell tolling near the city hall at 7 p.m. every evening, after which any 'non-Europeans' caught on the streets without an explanatory note would be charged and incarcerated for the night.

With the onset of a formal policy of racial segregation commonly known as Apartheid, during the mid-twentieth century this exclusionary system became further entrenched when the 'Homelands' policy was introduced, where workers and their extended families were expected to reside in designated areas, unless they had formal employment and were prepared to commute long distances to work every day. Their only alternative was to live in massive, single sex hostels situated in secured compounds, near to places of employment at mines, industrial areas or on the edges of cities.

As I spent much of my formative years during World War II in Thokoza hostel, this account is drawn from personal experience. My great aunt was instrumental in building the hostel during the late 1920s as a refuge for women and, as a Zulu linguist, subsequently served as a matron. My aunt followed and liberalized the institution by changing the name from 'Bantu Women's Hostel' to 'Thokosa' (Zulu for Happiness) after a democratic process.

There is very little written academic discourse about hostels, apart from a relatively obscure BA (Hons) thesis by Clarissa Dorothy Nelson, "Thokoza Hostel" submitted at the University of Natal in 1977. However, it says nothing about the interior design aspects. Harber and Associates submitted a report some years ago on the general topic.

These hostels were built almost invariably for men and have become a way of life, albeit abnormal. Men are crowded into large multi-storeyed dormitories around shared ablution blocks, with open, cold showers and rows of toilets. Food and liquor are available downstairs from informal stalls, with a beer hall for entertainment and womanizing. Most men acquire 'town wives' which is not necessarily frowned upon in a traditionally polygamous society. In most hostels separate quarters were originally built for traditional wives, who were allowed to stay over for a short while every year.

The background to hostels is deeply embedded in a huge number of socio-political issues. I have, for example, written extensively about how 'town wives' have led to the proliferation of HIV/AIDS back into rural areas. In remote Maputoland, for example, the infection rate amongst females is 47 %.

Ironically, since the new democratic government came into power, this hostel model has persisted, in spite of conversions of existing hostels to 'family units'. The explanation is simple: families lose their traditional access to land when members are not permanently resident.

The sleeping areas in hostels are drab, hall-like dormitories, furnished sparsely with up to 30 boldly numbered steel beds fitted with wire springs, covered by bare coir mattresses and including a basic shelf for storage. A worker is allocated a bed and use of water and electricity after paying a rental of less than US$ 10 a month. To reduce costs even further, or to help out a friend, this bed may be 'hot bedded' or shared by a night shift worker, thereby reducing costs further.

Male residents normally have no rights to the space below their beds. This space may be allocated during the course of the evening to ad hoc visitors or 'mechanics', as they are called by the administrative staff. These strangers normally sleep on corrugated cardboard and are only too grateful for a safe position away from the streets, although the invasion of privacy is beyond comprehension.

Since they have no permanent right to the spaces below bed level, men compensate for the faceless environment by expressing their individuality. Walls are decorated above their bedsteads with a rich array of posters, cut-outs, and drawings depicting political and religious allegiances, fantasy landscapes, official documents and photographs of soccer teams, family events, children and work places. A decorated steel trunk, tightly packed, is generally on the shelf to reinforce the connections to home and, if additional storage is needed, steel wire is strung across the room below the ceiling, to support hangers of ironed clothing, packed plastic bags and footwear. Bravado is the order of the day and the testosterone is palpable!

The few hostels for women are a direct contrast. This article focuses on the Thokoza (Happiness) hostel, which is well located right on the edge of the central business district of Durban. With over 5,000 residents per hectare, it is the most densely inhabited residential site in the city, crowded with women seeking independence and a safe haven from a male-dominated society. Street traders, dressmakers, domestic servants, 'washer women', cleaners, beaders, students and office workers are crowded into small bedrooms, each with three steel beds, lockers, and space for the most recent arrival to sleep on the floor.

The bedrooms are stacked four storeys high around the periphery of the site, with courtyards and ablution blocks between. At night the whole site is packed even more densely with itinerant women sleeping on plank platforms in corridors or on verandahs, to seek refuge from the streets. These are simply stacked up during the daytime.

Men are totally excluded.

The stark beds in their rooms are transformed with brightly coloured blankets to express their individuality. These are generally covered with clear plastic to deter bugs and avoid damp or damage, since this is their only space allocation, and the bed also serves as an eating table, a wash stand for relative privacy or space to earn a living. For example, it may support a dress-maker's hand-operated sewing machine.

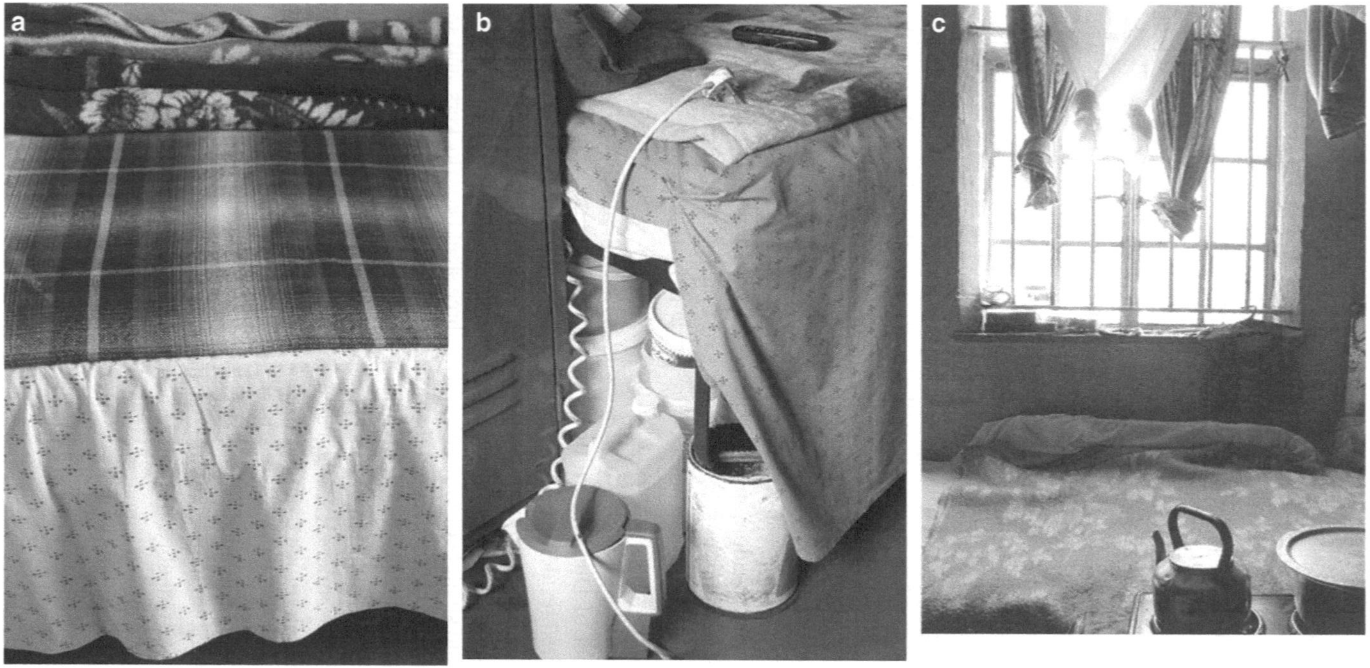

Fig. 6.1 **a** Makhanya from Mikhomazi, street trader, resident 50 years. **b** Thembeni Mncwabe from uMKmomaas, waitress, resident 11 years. **c** Mrs Nora Hlmabisa from Verulam, temporary domestic worker, resident 15 years. From left, Fig. 6.1a, b and c. When one lifts the neatly layered side drapes on the beds, slide away kitchens, neatly stacked clothing, footwear, cloth and special items for the next visit home are revealed. Peanut venders, for example, dry their groundnuts under their beds. Photographer Angela Buckland

It is interesting to note that whereas, in the men's quarters, senior men favour and are allocated corner beds, presumably for more privacy and wall space, women prefer the window bed in the middle, so that light can shine through the carefully arranged drapes and also illuminate a workspace.

It is immediately evident that women are spontaneous homemakers. Wall decorations are more demure and often colour-coordinated, with neatly stacked cloth, calendars, embroidery, alarm clocks and religious tracts, in contrast to the male environment.

The women's beds invariably stand on old 5 L paint tins. Some say this is to deter the Tokoloshe, a malevolent hairy manlike creature, but this is unlikely because people traditionally sleep on grass mats directly on the floor. The tins may deter cockroaches or even mice, but the real reason is probably very practical: storage space!

The lesson from the above is that if residents' needs had been anticipated or observed at the outset, beds could have had longer legs to facilitate storage, form a reasonable working height, or even more space for those who sleep under beds! Wall finishes should have been selected to encourage occupants to express their individuality by pinning up pictures, or to facilitate the fixing of ancillary units for more storage or displays. Tin trunks could be supported on brackets to free up shelf space.

Interior designers are traditionally image-makers for others with generous budgets. The evidence from South African hostels illustrates how residents have responded with expediency to facilitate bed-based livelihoods, and proclaims the importance of expressing their individuality and personal dignity.

Hostels have proven to be problematic since they are open to manipulation by political parties, but the most compelling argument against them is the spread of HIV/AIDS from 'town wives' taken back into the hinterland. On the other hand, high density inner city accommodation has merit. Hence, if similar institutions are to remain, the lessons set out above are only a starting point. The impact of design and especially no design has far reaching consequences. Therefore, Social Justice must be a compelling design imperative throughout.

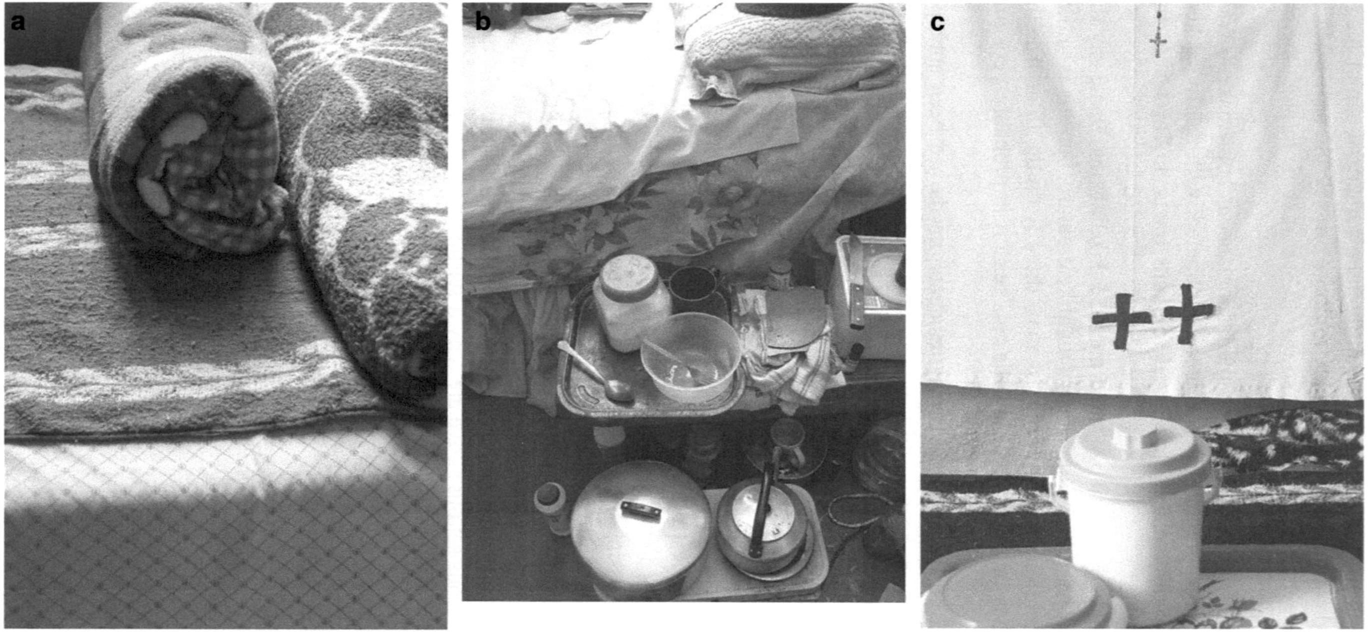

Fig. 6.2 **a** Makhosi Mthethwa from uMKmomaas, bead seller, resident 30 years. **b** Janet Mngoma from Groutville, cleaner, resident 11 years. **c** Zodwa Mary Jane from Matatiele, pensioner, resident 13 years. From left Fig. 6.2a, b and c. Raised bedstead to gain storage volume below. Photographer Angela Buckland

Chapter 7
Inside the Remote-Area Aboriginal House

Paul Memmott

Abstract This paper reports on the design requirements for Australian Aboriginal houses. Unfortunately, funding agencies and architects in the Indigenous housing sector consistently continue to provide houses to Indigenous people that are equipped for relatively small nuclear families, which results in a lack of fit between the housing designs and the Aboriginal domiciliary behaviours; yet government agencies are often guilty of imposing the former on the latter. Although this housing sector has frequently involved public servants, builders, engineers and architects since the 1960s, there has been negligible involvement by interior designers. Nevertheless, an understanding of the culturally distinct Aboriginal domiciliary behaviour should inform the process of the interior design of remote-area Aboriginal houses. There is potential for interior designers to expand their professional involvement in Aboriginal housing provision, provided that a sound and sensitive cross-cultural design methodology can be acquired and applied within the context of all-too-often constrained budgets and mainstreaming policies of government, which typically may mitigate against such involvement.

The structure of Aboriginal households is often quite different to Anglo-Australian households, and to design appropriate residential accommodation for Aboriginal people who have traditionally-oriented lifestyles, architects or interior designers must understand the nature of those lifestyles, particularly in the domiciliary context. Knowledge involves understanding those who have undergone cultural changes, including those in rural, urban and metropolitan settings, by helping to identify those aspects of their customary domiciliary behaviour that have been retained. Aboriginal kinship systems are very different to Anglo-Australian ones; they provide patterns of behaviour for many of life's situations, the patterns being represented or codified by the various types of relationship; these must be addressed through the spatial layout and occupancy.

Keywords Aboriginal · Interior · Domicile · Lifestyle customs · Design

P. Memmott (✉)
Aboriginal Environments Research Centre (AERC), School of Architecture and Institute for Social Science Research, University of Queensland, Brisbane, Australia
e-mail: p.memmott@uq.edu.au

D. Smith et al. (eds.), *Perspectives on Social Sustainability and Interior Architecture*,
DOI 10.1007/978-981-4585-39-2_7,

Introduction

Whereas the provision of design services for Australian Aboriginal houses since the 1960s has frequently involved public servants, builders, engineers and architects, there has been negligible involvement by interior designers. Nevertheless, an understanding of the culturally distinct Aboriginal domiciliary behaviour should inform the process of the interior design of remote-area Aboriginal houses. I have previously outlined a more systematic analysis of the behavioural themes in this paper, including a detailed list of ethnographic references underlying the reported findings (Memmott 2003, pp. 26–39). Elsewhere I have also described a body of architectural knowledge based on such understandings as the 'cultural design paradigm':

> … Its premise is that to competently design appropriate residential accommodation for Aboriginal people who have traditionally-oriented lifestyles, architects [or interior designers] must understand the nature of those lifestyles, particularly in the domiciliary context. This knowledge also increases understanding of the needs of groups who have undergone cultural changes, including those in rural, urban and metropolitan settings, by helping to identify those aspects of their customary domiciliary behaviour that have been retained. (Memmott and Go-Sam 2003, pp. 12–17)

This premise joins together with the required fiduciary duty of care of the professional designer towards their householder client when addressing their architectural needs, and with the Social Justice principle of avoiding indirect discrimination. Indirect discrimination, as it is construed in Australian law, can occur in Aboriginal service delivery when a set of conditions that are normally applied to all Australians with a level of compliance and acceptance, are transposed to Aboriginal Australians who, because of their cultural values and practices, cannot readily comply with or accept the conditions in the way that Anglo-Australians do, and thereby suffer unreasonable discomfort and stress. This typically occurs under mainstreaming policies when a government decides 'one size must fit all' (Memmott 1990). In Aboriginal housing delivery, questionable assumptions commonly built into government policy include the idea of the nuclear family being a norm, and the expectation that families will display a patterned distribution of domiciliary behaviour-types in conventional household spaces according to Anglo-Australian norms.

The Pervasive Influence of Aboriginal Kinship on Household Behaviour

Aboriginal kinship systems are very different to Anglo-Australian ones; they provide patterns of behaviour for many of life's situations, the patterns being represented or codified by the various types of relationship, such as father-son, mother's brother, sister's son and so on. The behaviours are both positive and negative; that is, a certain relationship may demand that the two persons concerned perform certain duties towards one another, for example, sharing of food resources. However, it may also prescribe that certain things be not done, for example, that no direct communication occur between certain relatives (Elkin 1938, p. 69). These behavioural rules manifest in the domiciliary spaces of remote Aboriginal communities.

Some of the rules and obligations based upon kinship will affect socio-spatial behaviour, or the actions of people in spatial relationships towards one another—either their relative spatial positions, their orientations, or the extent of their body contact—seen as an expression of their particular social relations. A new arrival in a domiciliary group can precipitate a shift in spatial behaviour, depending upon the particular kinship relationships between the newcomer and the existing members of the group. A special category of respect among certain tribal groups involves the rules of avoidance between men and their mothers-in-law, and between women and their adult brothers. These rules apply to small group behaviour where individuals are present in a single room of a house (for example, watching television), but also underpin how residents may be sub-divided into sub-groups occupying separate parts of the house.

Remote Aboriginal Household Structures

The structure of Aboriginal households is often quite different to Anglo-Australian households. Residential surveys of remote settlements throughout recent decades have indicated the persistence of customary domiciliary or household groups in both northern and central Australia, such as (a) the nuclear family, (b) the single men's group, and (c) the single women's group, as well as the emergence of other structurally diverse household forms that have arisen partly because of the cultural change effected in domestic economies (welfare benefits), social authority structures and residence in Western houses (Keys 1996, pp. 69–77, 1999, 2003, pp. 64–71).

In many cases we find several customary family units occupying a single house, each residing in a separate bedroom. Houses therefore do not necessarily correlate with single family units in contemporary Aboriginal societies. Aboriginal households tend to be larger and more complex, often including a number of family sub-groups who are interrelated within a lineage. Household sizes of six to twelve people are common, and much larger multi-generational ones of up to twenty members can be regularly encountered. Some of these large households have a matrifocal structure.

Remote Aboriginal families do not use house rooms in the same pattern as Anglo-Australians, with a clear division between the rooms of shared social use (living, dining, kitchen etcetera) and private bedrooms. In the largest Aboriginal households, it is normal to find each bedroom occupied by a family unit comprising, for example, a couple with infants, or a single parent with a child, or a group of single men or single women, or a grandparent with several infants or teenagers. Such a sub-unit would be considered a family unit in mainstream Australian society. In some cases these families are residing together because of a shortage of housing. However, in many other cases they may choose to reside in such large household groupings, in keeping with customary kinship obligations (which demand sharing and reciprocity).

In reality, these averages are even higher at particular times of the year because of visiting kin. According to Aboriginal kinship practices, the sharing of accommodation with certain relatives is a social responsibility. Visitors will usually stay at a relative's house for anywhere between a single night and a number of months. Indigenous household structures are therefore typically dynamic.

The Relationship Between Large Indigenous Households and Crowding

The social sciences have employed a stress model of 'crowding' for at least 40 years. This model holds that states of crowding involve high-density settings that generate certain stimuli. Such settings induce stress in the participants according to their values regarding the environmental acceptability and non-acceptability of these stimuli. In the case of remote Aboriginal groups, the stimulus that induces stress is often the presence of inappropriate categories of kin in too close a proximity. Such a model emphasizes the presence of perceived loss of personal control as a pre-requisite of crowding, and holds that the cause of such a loss of personal control may vary cross-culturally (such as perceived forms of stressful density), as will behavioural norms and rules for minimizing such stress. A coping mechanism may be used if one is available; a reactive behavioural component of the model. The values that are employed to evaluate the setting state (its stimuli), and to select an appropriate coping mechanism, vary cross-culturally. In terms of spatial scale, one can differentiate between room crowding, house crowding and neighbourhood crowding (Memmott 1991; Gifford 2007). All of these can occur in remote Aboriginal communities.

To minimize the stresses arising from high density living in remote Aboriginal households, a common coping mechanism is the purposefully arranged setting structured by the householders, achieved through rules governing the combinations of people allocated to living and sleeping spaces, which introduce what are perceived to be ordered and safe behavioural patterns. If, for example, a sub-group of unmarried women is allocated a room in a large household, their numbers are unlikely to be a concern and they will sleep within touching distance of one another. The arrangement of people in sleeping spaces thus occurs according to combinations based on age, gender, conjugal status and kin relationships. Despite being a large household it may not be regarded as overcrowded. If the core members of such a rule-governed household are stable, such households may endure for years (Memmott et al. 2012; Musharbash 2008).

One sub-group of householders—often including the senior householder—may sleep and live in the 'living room' of the house, irrespective of whether bedrooms are too small or too few. The room is furnished with mattresses on which people will sit or lie engaging in social discourse or sleep as they wish. This differs from the typical Australian living room, which often features a couch and a television, but which is seldom used as a nocturnal sleeping room.

A threshold of stress may arise, even for the rule-governed household, when the density increases to the point whereby there is no means of allocating sleeping space to persons without placing them in situations which compromise the need for respect among kin. Such a situation will bring stress, and emotional responses may include shame, jealousy, anger and violence. The household in this situation is generally crowded. The household can be severely disturbed through substance abuse by particular householders or by their visitors. In such a case, it will lose stability and may not survive (Memmott et al. 2012).

Designing Houses for Large Household Sizes with Culturally Specific Behaviours

Architectural design can either support or undermine the social sustainability of kinship. Since contemporary remote Aboriginal households are structurally variable, and frequently complex, the designer should first profile them before embarking upon the design of their houses.

The designer should not at first assume large households are crowded, although it will be found in certain cases that they are. In large households made up of a number of discrete sub-groups, between whom tensions may arise, consideration must be given to allowing each subgroup some spatial independence, and visual and aural privacy. Designers may need to provide sufficient toilets, bathrooms, and multi-purpose activity rooms for household sub-groups, as well as climatically appropriate design. Such multi-functional rooms are not confined to sleeping and dressing functions, but may also need to accommodate eating, storage of food and other possessions, watching television and DVDs, and listening to music. The members of household sub-units often wish to store all their personal items in their own rooms, which can include fridges, hunting gear, clothes and toiletries (Memmott et al. 2000, pp. 41, 46).

Sufficient spaces and workable room layouts are required in the house for individuals to congregate in small groups without compromising avoidance relationships. Feelings of tension and stress can be created by poorly designed room layouts and lines of vision, such as those that give unwanted visual surveillance of persons in avoidance categories approaching toilet doors (Fantin 2003).

Unfortunately, funding agencies and architects in the Indigenous housing sector consistently continue to provide houses to Indigenous people that are equipped for relatively small nuclear families (Memmott and Moran 2001). This results in a lack of fit between the housing designs and the Aboriginal domiciliary behaviours, yet government agencies are often guilty of imposing the former on the latter, which can be construed as a form of indirect discrimination, and one that threatens social sustainability. Designers thus need to understand the relevant socio-spatial behaviours enacted in remote Indigenous households (Memmott 2002).

Designing for Activities in Semi-Enclosed and Outdoor Spaces

Large households will overflow into semi-enclosed and outdoor spaces. Verandah spaces must accommodate both the householders and their daytime or overnight visiting kin. For example, there may be single gender groups who meet and play cards daily at a particular house.

Many people prefer to socialize and even sleep outside, unless this is made too disagreeable by mosquitoes, rain or intoxicated persons. It is important to note that some people will sleep on verandahs at night-time, whilst other people or households will only use verandahs if they are secured with screens for reasons of safety and privacy (Memmott et al. 2000).

Verandahs have to perform a number of seemingly conflicting climatic control functions—to provide maximum exposure to ventilation in hot-dry and hot-wet times, and to provide protection from wet weather. While it is often important to provide some form of privacy to verandah spaces (screens, slats, blinds), people also wish to have clear views; therefore verandah edges must be designed with the sometimes competing concerns of surveillance and privacy in mind. However, a balance is achievable with good design.

Conclusion

There is thus a conundrum in writing a piece on the interior living of remote Aboriginal families, since a distinct norm is to live outside of the house or in semi-enclosed spaces, where people can maintain wider social surveillance. In relatively fine weather, interior rooms may become relegated to the storage of possessions, and their significance is de-activated (Musharbash 2008). Nevertheless, interiors will be re-activated in inclement weather. There is thus potential scope for interior designers to expand their professional involvement in Aboriginal housing provision, provided that a sound and sensitive cross-cultural design methodology can be acquired and applied. Unfortunately the all-too-often constrained budgets and mainstreaming policies of government typically mitigate against such involvement.

References

Elkin, A. P. (1938). Kinship in South Australia part 2. *Oceania, 9*(1), 69.

Fantin, S. (2003).Yolngu cultural imperatives and housing design: Rumaru, Mirriri and Galka. In P. Memmott & C. Chambers (Eds.), *TAKE 2: Housing design in indigenous Australia* (pp. 72–79). Canberra: Royal Australian Institute of Architects.

Gifford, R. (2007). *Environmental psychology: Principles and practices*. Canada: Optimal Books.

Keys, C. (1996). Defining single Aboriginal women's needs in Central Australia: Dealing with issues of culture, gender and environment. *Architectural Theory Review, 1*(1), 69–77.

Keys, C. (1999). *The architectural implications of Warlpiri Jilimi.* (Unpublished doctoral thesis). The University of Queensland, Brisbane, Australia.

Keys, C. (2003). Housing design principles from a study of Warlpiri women's Jilimi. In P. Memmott & C. Chambers (Eds.), *TAKE 2: Housing design in indigenous Australia* (pp. 64–71). Canberra: Royal Australian Institute of Architects.

Memmott, P. (1990). *Racism, anti-discrimination and government service delivery in Aboriginal New South Wales*. Discussion paper for the NSW Anti-Discrimination Board, Attorney-General's Department, New South Wales. Aboriginal Data Archive, Department of Architecture, University of Sydney.

Memmott, P. (1991). *Humpy, house and tin shed: Aboriginal settlement history on the Darling River*. Sydney: Ian Buchan Fell Research Centre, University of Sydney.

Memmott, P. (2002). Sociospatial structures of Australian Aboriginal settlements. *Australian Aboriginal Studies, 1,* 67–86.

Memmott, P. (2003). Customary aboriginal behaviour patterns and housing design. In P. Memmott & C. Chambers (Eds.). *TAKE 2: Housing design in indigenous Australia* (pp. 26–39). Canberra: Royal Australian Institute of Architects.

Memmott, P., & Moran, M. (2001). *Indigenous settlements of Australia.* (Technical paper). Canberra: Environment Australia. Available at: http://www.au/soe/techpapers/indigenous/introduction.html.

Memmott, P., & Go-Sam, C. (2003). Synthesizing indigenous housing paradigm: An introduction to TAKE 2. In P. Memmott & C. Chambers (Eds.), *TAKE 2: Housing design in indigenous Australia* (pp. 12–17). Canberra: Royal Australian Institute of Architects.

Memmott, P., Long, S., Fantin, S., & Eckermann, K. (Eds.). (2000). *Post-occupancy evaluation of Aboriginal housing in the NT for IHANT: Social response component.* Darwin: The Architects Studio, P Memmott & Associates, Rawlinsons. Available at: http://www.aboriginalenvironments.com/docs/Memmott_Long_Fantin_Eckermann_2000.pdf.

Memmott, P., Birdsall-Jones, C., & Greenop, K. (2012). *Australian Indigenous house crowding*. Final Report 194 for the Australian Housing and Urban Research Institute, Melbourne. Available at: http://www.ahuri.edu.au/publications/projects/p20640.

Musharbash, Y. (2008). *Yuendumu everyday: Contemporary life in remote aboriginal Australia.* Canberra: Aboriginal Studies Press.

Chapter 8
Too 'Nice' to be Socially Responsible: Reflections of a Danish Interior Practitioner

Marianne Frandsen

Abstract This paper asks of the reader: why are others such as artists considered capable of doing work in the area of socially responsible design, that would normally be deemed the core business of interior architects, in Copenhagen? And in association, why do interior architects appear to be absent from this area of practice? Through Danish examples, the ability of the artist to advocate more freely for the marginalized groups in society as well as the general populus is illustrated. In Denmark, partly due to the welfare state, the design of social institutions, ranging from hospitals to kindergartens, has always been a public matter. In recent years some of the public institutions that have special social responsibilities have asked artists to develop interior architecture projects. As a result of the artists' involvement and outcomes, which often include a political statement, public attention has been aroused through the media. Development of the New University Hospital (DNU) is based on the premise that better and more beautiful surroundings created with art and design foster faster healing, and an arts strategy was formulated to facilitate this. It is proposed that it is important to incorporate social layers as well as the usual physical ones in the research and design for a project. Opportunities exist to apply the healing, wellbeing ethics of Evidence Based Design philosophy to the creation of beautiful, stress-free spaces in our everyday interiors. Interior designers need to be involved in this important area of work across the breadth of projects from homeless people's shelters to large public buildings.

Keywords Interior designers · Artists · Advocacy · Evidence based design · Design

Introduction

Why are others such as artists considered capable of doing the kind of work that would normally be deemed the core business of interior architects in the area of socially responsible design, in Copenhagen? And in association, why do interior architects appear to be absent from this area of practice?

M. Frandsen (✉)
Designer, honorary member of Danish Designers, Copenhagen, Denmark
e-mail: frandsen.haagensen@gmail.com

D. Smith et al. (eds.), *Perspectives on Social Sustainability and Interior Architecture*,
DOI 10.1007/978-981-4585-39-2_8, © Springer Science+Business Media Singapore 2014

It is usual for designers, artists and architects to work across discipline boundaries to engage with interiors. Two examples of interior spaces designed by artists are noted to demonstrate this. These are Piet Mondrian's Atelier in the 1920s and Donald Judd's New York and Marfa interiors in the 1960s and 1970s. Although a common distinction made between design and art has been that design must have a function while the arts do not normally have this quality, both Mondrian and Judd have challenged this assumption. By working in three dimensions, their art has evolved to have a simple function; that is, one can live in the resulting spaces. In these instances, the interior spaces could simply be considered a three-dimensional extension of their art form.

Over history, fine artists have been part of the revolutionary thinking of any given period. During the last 20–30 years, such artists have become much more aggressive in accessing the three-dimensional world of space by engaging in the creation of installations. Through this trend, they have eventually taken over other areas that traditionally were occupied by architects and interior architects.

What is it in our period that gives artists the urge to design interior spaces? In relation to Social Justice projects, why do they choose to give their qualified bid for the often difficult, socially impacted spaces such as shelters and prisons? The obvious answer would include social indignation and a professional ethical attitude, which leads to the arts becoming an outstretched hand, and often a spectacular statement about social injustice as well.

It would seem pertinent to reflect here on a very simple and banal aspect of the history of interior architecture, arts and architecture. Artists and architects have a history built over hundreds of years and as a result they rarely have to explain what their professions do. In contrast, interior architecture is still emerging in modern times. While interior architects have fought over the last 60 years to explain themselves and the nature of their profession, artists have been able to continue their tradition of being part of the commentary on this period's serious problems in society. Thus, artists have been able to invade the tenuous territory of designing interior spaces, often in enviable ways.

A Danish Perspective

In Denmark, partly due to the welfare state, the design of social institutions, ranging from hospitals to kindergartens, has always been a public matter. In recent years some of the public institutions with special social responsibilities have asked artists to develop interior architecture projects. As a result of the artists' involvement and outcomes, which often include a political statement, public attention has been aroused through the media. To demonstrate this phenomenon, a case study involving an artist will be discussed, followed by another that integrates interior architecture practice.

Case Study One: Home for Men

One case, Home for Men, is a restored/redesigned shelter for homeless men in the area close to the Central Station of Copenhagen. As in many cities, the buildings around the main station were built in the last century for people coming to the city for jobs in new industries. In less fortunate times, the area became a haven for the jobless, alcoholics and men living in the streets. Today the area has been renewed and thousands of people arrive every day by trains and buses from the Nordic countries and Europe. As a result, it is the most cosmopolitan area in Denmark. The drug culture, a volatile mix of races, and prostitution exist side by side with chic cafés, ethnic restaurants, international kiosks and residential areas, to form a diverse hub.

The Home for Men was established in 1906, mainly through the support of Christian and other religious organizations. In the 1960s, the home became the responsibility of the official Danish welfare policy and is funded by the municipality of Copenhagen.

The Home's design has changed over the years, with the latest major change occurring in 2002. The institution commissioned two artists—Thomas Poulsen, who is known as FOS, (Poulsen 2011) and Kenneth Balfelt (Balfeldt 2011). Working in the area of social design, the definition and meaning of their practice focuses on ways in which arts and design may be used to change the idea of a social space or room. Their basic premise is that a space or room consists of both a social dimension and a physical reality. The artists view the social space as a tangible outcome that can be formed in the same way as a physical space can be created. FOS stated that 'all the social relations/layers that a society is built of can be considered as physical as steel, stone et cetera and can therefore be seen as elements in a design process' (Poulsen 2008, p. 24).

Many of the design themes were basic. For instance, conflict prevention, horizontal relations between users and social workers and abolishing the traditional patriarchal patterns in the institution were all addressed. The two existing and strongest patriarchal symbols, the reception counter, looking like a border control post, and the attendant's office, were removed. The entrance, which had earlier been placed 'discreetly' as a dark gate in the side street became visible at the street corner and was designed as a solid oak door with ornaments and a copper canopy that provides shelter from the weather.

The artists' idea was to also reference homeless lifestyles and develop appropriate associations. The former reception is now an observation room—also used for sleeping at night—designed as a train compartment, in reference to the homeless who spend many nights in trains. In this world of street life, meetings and meeting rooms symbolize power. In contrast, a camper van, often a homeless person's dream, is included as a room for an informal dialogue (Fig. 8.1). 'Flake' or 'chill-out' chairs are placed in a café-like room. Durable materials like solid wood, terrazzo, tiles and metals, as well as colour, are used to create a safe base in the users' unstable lives.

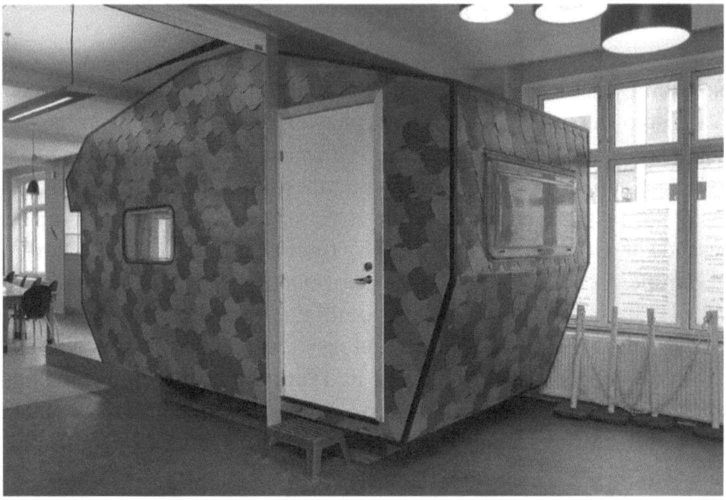

Fig. 8.1 Camper for Dialogue; photographer Dorte Krogh

The pattern of users is changing rapidly. The diverse users of the shelter include immigrants, refugees, drug addicts, prostitutes and trafficked women who are lured by promises of normal jobs upon arrival, and are then traded to pimps, who then take their passports. They are traded mostly from Africa or Eastern Europe and therefore they are no longer restricted to middle-aged male alcoholics, who appear to have been the dominant users in the past (Olsen 2007; Vibæk and Døssing 2004).

FOS has defined his work as the genre of social design. By navigating outside the arts institutions, he wants to give experiences to people who are not normally exposed to fine arts. He is conscious of his role as an artist: 'in a society marked by rising irresponsibility, I have the opinion that arts has an increasing responsibility to give alternatives to the society models given us today' (Poulsen 2010). FOS is not wedded to a particular moral or ethical principle but believes that 'an artistic grasp can create differences in society' (Poulsen 2010).

When interviewed recently about working with the shelter, he revealed that it had been a shock, and he found it hard to be confronted with the addicts' milieu;

> The users are symptoms of what our society rejects, it tells more about us than about them. During the last six years the pressure on the shelter has increased—it is like walking in an open wound. They are lying on top of each other fixing. But all want a nice façade without syringes .... (Poulsen 2010)

The artist's provocative involvement in social spaces became a resistance or guerilla design aimed to challenge an existing anti-humane policy in a rich welfare state.

Case Study Two: New University Hospital

A contrasting example is a project for The National Arts Foundation. Early in 2010 it was decided that designers should participate in developing an arts strategy for

Fig. 8.2 The Reading Room; installation by AVPD; photographer Dorte Krogh

the New University Hospital (DNU), Aarhus, Denmark—the first of five huge new hospitals in Denmark (DNU 2011). The project involves social responsibility, as the hospital's basic design strategy is Evidence Based Design (EBD), and has the premise that better and more beautiful surroundings created with art and design foster faster healing (Fig. 8.2).

A working group was established to formulate an arts strategy for DNU. For the first time, members from the Danish design world were included to work with artists—a photographer, Dorte Krogh, and myself. Although we had unsuccessfully tried to change the name of the strategy to 'The arts and design strategy', we have found many ways of planning for the integration of arts and design in the future design of the large hospital. The Reading Room illustrates the strategy and the collaboration between artists and designers (Fig. 8.2). In the arts strategy, this spatial construction by Aslak Vibæk and Peter Døssing is an example of an installation where the viewer's normal perception of space is challenged (Kunsthallen and Nicolaj n.d.).

The strategy is a tool for the future use of politicians, planners, interior architects and architects, for clients among regional authorities, as well as for incoming curators. The introduction to the strategy reflects the tone of the document:

> One never really forgot that the building was there for the sake of the patient. But rational professional practice, efficient organising, investments, calculations and technological instruments make their demands. Therefore it is easy to forget that the building is also a

> nurse, one who communicates directly with the patient and without words. Moreover the building is often the most covering expression for the ideology that justifies care for the distressed. (Eriksen 2008, p. 181)

Conclusion

As Eriksen says in the quote above, ' … the building is a nurse, one who communicates directly with the patient and without words.' This statement underlines so brilliantly the essence of what we are working towards. As interior architects, our profession should include social indignation as part of our work.

To incorporate social layers as well as the usual physical ones, as highlighted by FOS, is relevant within both educational curricula and in the research prior to a project. It can be a new means by which to cooperate with clients. During my work with design thinking for future hospitals, I have wondered why the healing effect of EBD seems to have been limited to hospitals. It seems obvious that opportunities exist to apply the healing, wellbeing ethics of EBD philosophy to the creation of beautiful, stress-free spaces in our everyday interiors.

References

Balfeldt, K. (2011). http://www.a-r-d.org/.

DNU (The New University Hospital). (n.d.). *Det Ny Universitetshospital i Åarhus*. http://dnu.rm.dk/.

Eriksen, T. (2008). Mer enn det du ser. In M. Paasche & A. Hovind (Eds.), *Akershus University Hospital* (p. 181). Oslo,Norway.

Kunsthallen & Nicolaj. (n.d.). Retrieved from http://www.kunsten.nu/artikler/artikel.php?AVPD%2C+Kunsthallen+Nicolaj+.

Olsen, R. (2007). Nordisk sosialt arbeit. *Mændenes Hjem, 2,* 148–160. Retrieved from http://maendeneshjem.dk.

Poulsen, T. (FOS) (2006). http://www.kunst.dk/billedkunst/omtilskud/hvemharfaaettilskud/2006/treaarigearbejdsstipendier2006.

Poulsen, T. (FOS). (2008). *Hus Forbi,* (74), 24.

Poulsen, T. (FOS). (2011). Retrieved from http://www.socialdesign.dk/.

Poulsen, T. (FOS). (n.d.). Interview. AOK (Alt om København). Retrieved from http://www.aok.dk/byliv/mit-koebenhavn-fos. Accessed 16 Aug 2010.

Vibæk, A. & Døssing, P. (2004). Reading Room, Gallery Kirkhoff. Retrieved from http://www.avpd.net/pdf.PSS-ReadingRoom.

Part III
Cultural Heritage and Interior Architecture

Cultural heritage is the legacy of the tangible and intangible aspects that are passed from one generation to the next because they are perceived to have value to a society or a community. It includes, but is not confined to, objects, artefacts, ephemera and both the natural and built environments.

The following two chapters explore the notion of social sustainability through the cultural heritage of the built environment, ranging from the fine-grained domestic Victorian and Modernist interior through to the gritty large-scale environment of an industrial railway workshop. While the scale, range and functions of these places vary enormously, they are recognized as historical records of lived experience and identity. Their value to local and broader communities as the physical manifestations of cultural diversity and social evolution is recognized; and the retention and ongoing use, either for their original purpose or through adaptive re-use, is what makes them sustainable.

Such places are not necessarily pretty places, or reminders of a positive past—prisons and factories, for example, may be valued because they are a tangible reminder of a set of social relations that have evolved over time, and the memories associated with those places are embedded in their physical fabric.

> Places of cultural significance enrich people's lives, often providing a deep and inspirational sense of connection to community and landscape, to the past and to lived experiences .... Places of cultural significance reflect the diversity of our communities, telling us about who we are and the past that has formed us .... They are irreplaceable and precious. (International Council of Monuments and Sites Inc. 2000, p. 1)

References

International Council of Monuments and Sites Inc. (ICOMOS). (2000). *The Burra Charter*. Australia: ICOMOS. http://australia.icomos.org/wp-content/uploads/BURRA_CHARTER.pdf.

Chapter 9
Cultural Heritage Practice Through Socially Sustainable, Adaptive Re-use of Industrial Buildings: A Western Australian Narrative

Priya Metcalfe

Abstract Conservation and heritage practice in Western Australia has been, and continues to be, primarily concerned with the recognition and retention of places that have cultural heritage significance on a local, national and international level. Their value lies in a desire by the community to preserve important buildings, their sites and their interiors that speak of the evolution of architectural styles and practices, provide evidence of the layers of human modification of the natural environment, and contribute to the richness of Western Australia's built and social history. Their ongoing use, either for their originally intended purpose or through adaptation for new use, is evidence of the sustainability of such places.

One such place is the Midland Railway Workshops in Midland: a large complex of buildings, ancillary spaces, machinery and equipment, associated with the production and maintenance of rolling stock for the West Australian Government Railways system, which was vital to the ongoing growth and development of the State. The workshops were constructed from 1904 and remained in consistent use until the site's closure in 1994. The site is currently undergoing development for a variety of uses.

A significant outcome of the changes in function of the Railway Workshops buildings is the gentrification of what was a noisy, crowded, dirty and, in some cases, dangerous industrial site.

This chapter discusses this compromise as an alternative to the deterioration of places through lack of continued use, and argues that to do other than modify the buildings and their context would likely mean the permanent loss of a culturally significant site, and therefore the loss of its values, including its social sustainability to the local Midland community specifically, and the wider community of the state and nation generally.

Keywords Conservation · Heritage · Adaptive re-use · Midland Railway Workshops · Cultural heritage significance · Social sustainability · Heritage values

---

P. Metcalfe (✉)
Priya Metcalfe Conservation and Heritage Consultant, Perth, Australia
e-mail: p.metcalfe@bigpond.com

D. Smith et al. (eds.), *Perspectives on Social Sustainability and Interior Architecture*,
DOI 10.1007/978-981-4585-39-2_9, © Springer Science+Business Media Singapore 2014

Introduction

Conservation and heritage practice in Western Australia has been, and continues to be, primarily concerned with the recognition and retention of places that have cultural heritage significance on a local, national and international level. These places are embodied in memory and identity, and form part of the history of European occupation of the State. Their value lies in a desire by the community to preserve important buildings, their sites and their interiors that speak of the evolution of architectural styles and practices, provide evidence of the layers of human modification of the natural environment, and contribute to the richness of Western Australia's built and social history. Their ongoing use, either for their originally intended purpose or through adaptation for new use, is evidence of the sustainability of such places, whether that sustainability is social, economic or ecological.

One such place is the Midland Railway Workshops in Midland, east of the Perth city centre on the Helena River: a large complex of buildings, ancillary spaces, machinery and equipment, associated with the production and maintenance of rolling stock for the West Australian Government Railways system, which was vital to the ongoing growth and development of the State. The workshops were constructed from 1904 and remained in consistent use until the site's closure in 1994. At its peak during the inter-war period, the Workshops employed 40 % of the manufacturing workforce of Western Australia, with up to 3,500 highly skilled and diverse employees, and boasted the largest concentration of blue-collar workers in the State (Moredoundt 2009, p. 4).

The cultural heritage significance of the Midland Railway Workshops was formally recognized by the Heritage Council of Western Australia (HCWA) when the site and many of its buildings were listed as a permanent entry on the HCWA Register of Heritage Places in 2008. The site was also classified by the National Trust in 1999 and placed on the Register of the National Estate in 1980.

The Midland Railway Workshops HCWA Register documentation, Statement of Significance, is as follows:

Midland Railway Workshops, comprising buildings, plant and equipment, roads, railway tracks, open spaces and landscape elements, has cultural heritage significance for the following reasons:

- the place is an intact example of a major early twentieth-century railway workshop
- the place was the most substantial industrial complex established by the West Australian Government at the period around the turn of the century and both the structures and the organization of the industrial processes on the site involved extensive planning on a very large scale
- the place contained the main workshops serving the West Australian Government Railways from 1904 until 1994 and played a major role in the economy, development and daily life of the State for almost a century
- the place contains a number of significant buildings including the original brick industrial buildings built in 1904 and extended c.1912, which comprise the most important group of early twentieth-century brick industrial buildings in Western Australia

- the place contains a wide range of significant machinery much of which remains operable, as well as collections of associated tools, equipment, furniture fittings and documents
- the place contains a number of significant site features associated with either the industrial process or the workforce associated with the site, including the War Memorial erected after World War I
- the place was a major factor contributing to the development of Midland Junction, through the employment of workers in the area
- the place was a major employer and played a significant role in the training of apprentices for the various trades and professions represented on the site. (HCWA Register Documentation, pp. 2–3)

The HCWA assessment also identified the place as having aesthetic, historic, scientific and social value.

The site is currently undergoing development for a variety of uses of existing buildings, remediation works and construction of both commercial and residential facilities. As a result, what was once a cohesive industrial site with 'fundamentally one owner and occupant' will become a place that will have 'numerous occupants and possibly many owners or buildings with multiple tenants' (Hocking Planning & Architecture 2009, p. 4). The consequence will be a shift in the functional, historical and social relationships between the buildings and their users (Hocking Planning & Architecture 1995, p. 3). A significant outcome of the changes in function of the Railway Workshops buildings is the gentrification of what was a noisy, crowded, dirty and, in some cases, dangerous industrial site, and the consequent potential alienation of the primarily working-class occupants of the workshops, with most development geared towards tertiary-educated and medium to high income earners. If one accepts that an integral element of social sustainability includes the retention of a place's value to the demographic of its original occupants and users, the shift to a new use means that the social sustainability of the place to the primarily blue-collar workforce, who used and valued the workshops is, of necessity, compromised.

This chapter discusses this compromise as an alternative to the deterioration of places through lack of continued use, and argues that to do other than modify the buildings and their context would likely mean the permanent loss of a culturally significant site, and therefore the loss of its values, including its social sustainability to the local Midland community specifically, and the wider community of the state and nation generally.

The interiors of these places are of particular interest, since they are the loci of either existing or potential occupants, and their spatial planning, design and detail represent the range of original users from the apprentices and teachers in the Railway Institute, to the workers who created the finely handcrafted timber patterns for parts and machinery in the Pattern Shop, and the tradesmen working on the construction, repair and maintenance of carriages and wagons and associated machinery in the three main Blocks.

Within conservation and heritage practice, the adaptive re-use of a significant site such as the Midland Railway Workshops is typically concerned with conserving the context and the external form and fabric of existing buildings, and in some cases removing accretions that have little value or have a negative impact on the values of places. This means that the exteriors will ideally largely appear as they were originally constructed, and are contextually unchanged as little as possible. In some cases extensions have been made, but these are generally lighter in construction, and comply with the tenets of the Australia International Council of Monuments and Sites (ICOMOS) Burra Charter (specific articles from this document are included later), in that they are intended to be sympathetic in terms of their bulk, form, scale, character, colour, texture and material, and are readily identifiable as new work.

The interiors of these places, however, offer scope for interpretation, adaptation and spatial modification to accommodate new uses while retaining the values of the interiors. This allows the opportunity for practitioners specializing in historic interiors to put into practice a number of strategies that retain and reveal significance, while also proposing imaginative solutions for a prospective future use—one that assures that a place is socially sustainable through ongoing occupation and, in some cases, access to the broader community, including those who have an association with the site related to its original function.

This raises the question of what appears to be predominantly a lack of involvement in places of cultural heritage significance by graduates of the Interior Architecture program at Curtin University, although increasingly, final year undergraduates are proposing dissertation projects within heritage buildings. Currently, conservation and adaptation of these sites is generally managed by architectural practices and rarely includes the expertise of practitioners of interior architecture, although the Railway Institute is an exception on the Workshops.

Five case study buildings at the Midland Railway Workshops were selected for critique as individual places within a broader context because they represent various forms of conservation and heritage practice, including restoration, reconstruction, adaptation and interpretation, and all are examples of attempts at socially sustainable practice through either current or proposed re-use. They are: Blocks One, Two and Three (1904, 1912), the Railway Institute (1912), and the Pattern Shop (1904) (Fig. 9.1).

Within the context of the Railway Workshops site all five buildings are identified by HCWA as being of exceptional significance both locally and nationally. This level of significance imports substantial constraints on modification of existing internal and external fabric and context. In order to advise current and future users of these places, all have had conservation plans prepared as guiding documents for retention of their Cultural Heritage significance under an overarching conservation plan for the site as a whole. The Pattern Shop was adapted and extended in 2008 and currently houses an atelier of furniture designers, fine wood craftspeople and jewelers. Block One remains largely unused, although plans are underway for its conversion into a medical general practice super clinic. Blocks Two and Three are unoccupied, and the Railway Institute was restored and converted in 2002 into the main administration building for the Midland Redevelopment Authority (MRA), which is responsible for overseeing the conservation and development of the site. Both the Pattern Shop and Block Two were the sites of speculative propositions by Curtin Interior Architecture students in 2005 and 2007 respectively, suggesting

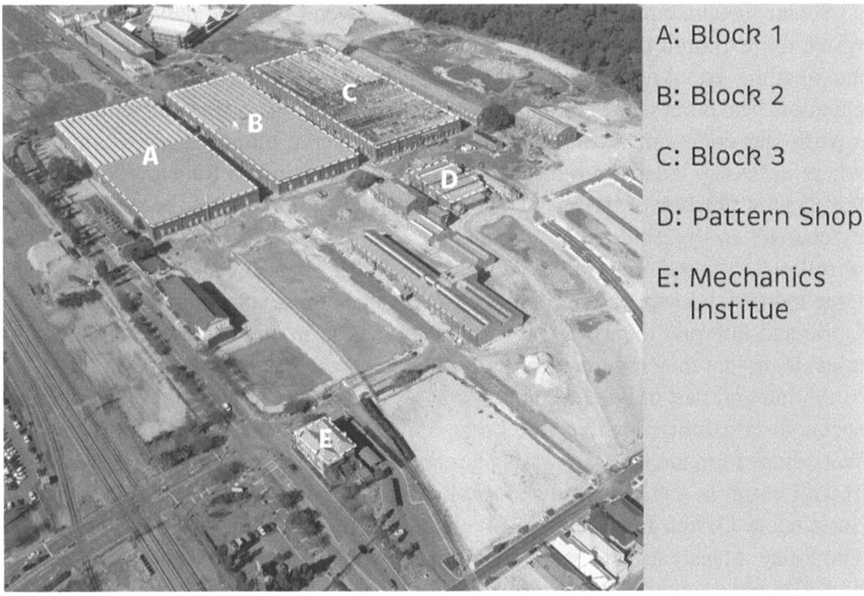

Fig. 9.1 Aerial view of the Midland Railway Workshops showing the location of the case study buildings. (Image courtesy of the Midland Redevelopment Authority)

solutions that, while realizable, and based on Burra Charter principles, were primarily concerned with a theoretical approach to interpretation and adaptive re-use. Both were concerned with the social sustainability of the sites, as discussed later in 'Current uses and sustainable future prospects'.

Cultural Heritage and Social Sustainability

The term 'sustainability' in the built environment is generally considered in relation to the 'triple bottom line' of outcomes in ecology, economy and society. Most frequently in recent years, ecologically sustainable design, whether in heritage buildings or new works—expressed out of concern for diminishing resources and a desire to avoid environmentally damaging materials and practices—has been a driving force in project development.

The economic benefits of retention and adaptation of existing buildings in preference to demolition are becoming more widely acknowledged. Recognition of the value of embodied energy and the cost of replacement of existing fabric are closely aligned to the concept of environmentally sustainable design. While both of these forms of sustainability are integral to the conservation and adaptive re-use of the Midland Railway Workshops, and were an overt part of the agenda of the MRA's Midland Metro Concept Plan in 2010 (Midland Redevelopment Authority 2010), it is the notion of social sustainability that is discussed in this chapter.

Social sustainability can be seen as having several possible definitions, two of which are explored here. One is the social values that are placed in the retention and ongoing use of places and their settings as part of the continuum of human occupation. The other is the recognition of the embedded histories and memories that a particular community values and wishes to preserve through conservation and, where necessary, adaptive re-use. In the case of the Midland Railway Workshops, the original function of the place, its buildings and equipment, has been superseded by changes in the technologies associated with the manufacture and maintenance of rolling stock and other railway machinery. Their retention, and their setting as a large industrial complex, therefore require changes in use that conserve significant fabric and interpret their original form and function in ways that are socially sustainable, in that they remain meaningful to past, present and future users.

An integral part of determining the values of a place, and therefore its potential social sustainability, involves the community to whom it may be important. The Workshops form an integral part of a larger redevelopment program for the Midland Town Centre. In 1997 the Midland Revitalisation Charette was led by Ecologically Sustainable Design Pty Ltd to resolve some of the planning issues of the center. The 5-day intensive workshop resulted in the Workshops precinct being identified by stakeholders as the key site for revitalization, and led to the formation of the Midland Redevelopment Authority in 1999. One indication of the significance of the place to the local community was the construction of a 'Workers' Wall', laid with bricks purchased by past workers and their families, each inscribed with their names.

The conservation and adaptive re-use of the Midland Railway Workshops is informed generally by Burra Charter's guidelines for restoration, conservation and adaptive re-use of culturally significant sites in the form of articles. Any changes to places of cultural heritage significance are principally guided by the Burra Charter articles, some of which are as follows:

Article 8. Setting

Conservation requires the retention of an appropriate visual setting and other relationships that contribute to the cultural significance of the place. New construction, demolition, intrusions or other changes which would adversely affect the setting or relationships are not appropriate.

Explanatory notes: Aspects of visual setting may include use, siting, bulk, form scale, character color, texture and materials.

Article 15. Change

15.1 Change may be necessary to retain cultural significance, but is undesirable where it reduces cultural significance. The amount of change to a place should be guided by the cultural significance of the place and its appropriate interpretation.

Explanatory Notes: When change is being considered, a range of options should be explored to seek the option which minimises the reduction of cultural significance.

15.2 Changes which reduce cultural significance should be reversible, and be reversed when circumstances permit.

Explanatory Notes: Reversible changes should be considered temporary. Non-reversible change should only be used as a last resort and should not prevent future conservation action.

Article 20. Reconstruction
20.1 Reconstruction is appropriate only where a place is incomplete through damage or alteration, and only where there is sufficient evidence to reproduce an earlier state of the fabric. In rare cases, reconstruction may also be appropriate as part of a use or practice that retains the cultural significance of the place.
20.2 Reconstruction should be identifiable on close inspection or through additional interpretation.

Article 21. Adaptation
21.1 Adaptation is acceptable only where the adaptation has minimal impact on the cultural significance of the place.
21.2 Adaptation should involve minimal change to significant fabric, achieved only after considering alternatives.

Article 22. New Work
Article 22.1 New work such as additions to the place may be acceptable where it does not distort or obscure the cultural significance of the place, or detract from its interpretation and appreciation.
Explanatory notes: New work may be sympathetic if its siting, bulk, form, scale, character, colour, texture and material are similar to the existing fabric, but imitation should be avoided.
22.2 New work should be readily identifiable as such. (ICOMOS 2000, pp. 4–7)

Each of the buildings is the subject of a conservation plan, concerned with conserving and revealing the values of the places and their context, based on a series of policies specific to that building that detail constraints and opportunities for future use.

Midland Railway Workshops: Construction to Closure

The Swan Valley region and Guildford, north-east of Perth, were settled following the establishment of the Swan River Colony in 1829. The selected site of the future Railway Workshops at East Guildford (later renamed Midland Junction, and finally Midland) was originally part of James Stirling's 4,000-acre estate (Hocking Planning & Architecture 1995, p. 22). Throughout the nineteenth century the Swan Valley region slowly developed, but it was not until the completion of the Eastern Railway from Fremantle to Guildford in 1881 that the area began to be more densely settled, followed by a population and resources boom from the late 1880s in the eastern goldfields. The Midland Railway Company of Western Australia established a workshop at 'a site about three miles east of Guildford, near the junction of the Upper Swan and York Roads' (Hocking Planning & Architecture 1995, p. 12). This privately-owned company was constructing a rail line to Geraldton from Guildford and workers established a tent village that soon resulted in what was named Midland Junction, derived from the connection of the Midland (privately owned) and Eastern (Government owned) Railways. In the years following the establishment of the area, however, the population of Midland Junction declined. In spite of the announcement in 1900 that the government would be constructing railway workshops to replace the inadequate facilities in Fremantle, south of Perth, the population of Midland remained low, with a commercial centre exceeding the needs of locals (HCWA 1992b, p. 6). It took some decades for the Midland population to grow and the settlement to develop, and the Workshops were significant in slowly attracting a workforce to the district.

Blocks One, Two and Three (1904, 1912)

The Workshops were centred on the three main masonry buildings, Blocks One, Two and Three. Each is a very large, single-storey, steel-framed late Victorian industrial building of locally-produced face English bond brickwork, with a corrugated-clad, saw-tooth roof, and together they form the visual focus of the Midland Railway Workshops site. All three are approximately 180 m long and 73 m wide, after extensions to each were completed to the east in 1912, effectively doubling their length. The Blocks were an integral part of the operations of the site, being used for the construction, maintenance and repair of rolling stock. Block One housed the carriage and wagon shop and the sawmill, Block Two was the boiler and blacksmith's shop, and Block Three was the machine and fitting and erecting shop. Locomotives entered the buildings through large doors in their eastern ends on tracks that ran through the interior. Openings between the Blocks meant that trains could be rotated within one building, using large roof-mounted cranes, and transported across to the next to the specialist trades housed there (Hocking Planning & Architecture 1995, p. 26).

Internally each Block is a single open space with high roofs at approximately 7.5 m from the bottom of trusses, making them of considerable scale. The Blocks have their interiors divided into bays, with the roof structure supported on a series of steel columns approximately 5 m apart. A significant feature of the interiors is the system of large overhead gantries and cranes that was an important part of the Workshops, essential for their operation, and influential in the design of most of the buildings. They were operated by a driver sitting in a cabin mounted on the crane, and were used to move machinery and locomotives within and through the buildings (Figs. 9.2 and 9.3).

The 1940s was a period of considerable change in the workshops due to the demands of World War II. Not only were the staff numbers reduced due to the recruitment of workers into the services, but the workshops were responsible for the manufacture of munitions. A number of buildings were extended and new buildings constructed to enable the workshops to fulfil contracts for munitions supply. One of the most significant aspects of this work was the introduction of women to the workforce in the years 1943–1945. A separate Shell Annexe was built on the east-

Fig. 9.2 Blocks One (*left*) and Two (*right*). (Photographer Priya Metcalfe)

Fig. 9.3 Interior Block One. (Photographer Priya Metcalfe)

ern side of Block Three for the munitions work (this has since been demolished). Other buildings were altered specifically to cater for the increase in women in the workforce. The areas that had the most significant increase in women, apart from the munitions work in Block Three, were the canteen, stores and the administration offices (Batterham 2006).

One of the principal functions of the Midland Railway Workshops was apprentice training for all trades (Hocking Planning & Architecture 1995, p. 33). Edward (Ted) King, turner and iron machinist, 1953–1959, described the culture of the workplace:

> New apprentices were considered fair game for tradesmen and older apprentices and there were all sorts of tricks played on them, like being sent for a long weight (wait), a left-handed hammer, a bolt hole, a sky hook, a right-handed drill and various other time wasting errands that usually entailed going from place to place around the shop. (Murdoch University n.d.)

Bob Wells, former car and wagon builder, also recalled pranks played on apprentices and the reason for them:

> It wasn't so much initiation; it was more a transference from their life as a school kid to the life on the shop floor… For the first three months the apprentice stayed with the one tradesman for three months at a time. The first three months was really conditioning a person to the idea of working, commencing work at half-past-seven and knocking off at quarter-past-four and being accountable for what you do in that entire period of time. Now, the practices of sending people for left-handed screwdrivers, or for left-handed spanners, or for self-tightening nuts, or for a box of holes were really to cause the apprentice to stop and think about what it was that he was being asked to do. It was part of the gathering of responsibility for where he was at. It wasn't a prank done for any mean purpose; it was a question of getting him to think of what was being asked for. (Murdoch University n.d.)

In spite of the apparent harshness of treatment of the apprentices, employment at the workshops was valued by both its apprentices and senior employees, with a substantial number of new workers following in their fathers' footsteps—in some cases, three generations being employed in trades. This handing down of both work and the skills associated with particular trades is evidence of the social value placed on the Railway Workshops as a primary employer of local workers. It is the continued sustainability of these social values, in spite of the redundancy of the functions of buildings such as the three Blocks, that has been a primary driver of the retention and adaptation of the place.

While much of the apprenticeship training was work-based throughout the Workshops site, there were also technical studies that were included in the education of the workforce, and much of this took place in the Railway Institute.

The Railway Institute (1913–1914)

Midland Railway Institute is a two-storey, brick and iron building located on the north-western perimeter of the site. It is a fine example of Federation Free architectural style, and is one of the buildings that contribute to the street presence of the Midland Railway Workshops. It represented the role of technical education for apprentices in various trades and professions in the Workshops. It features traditional construction materials and design techniques, and has a symmetrical façade with greater attention to detailing along its eastern and northern (public) façades (McAllister & Heritage and Conservation Professionals 2009, p. 71). Original timber doors and windows remain.

The place is valued both in the context of the Midland Railway Workshops and in its own right, particularly to past teachers and apprentices. Internally the building is well proportioned around a central turned jarrah staircase that gives a sense of openness and light, which is the result of careful planning. The quality of timber joinery, especially the staircase and fireplace surrounds, contributes to the aesthetic quality of the place in a primarily industrial setting (McAllister & Heritage and Conservation Professionals 2009, p. 69).

The two-storey building as originally constructed was designed to accommodate the Institute on the ground floor, and the Technical School on the first floor. The works were supervized by Mr. Logan, of the Ways and Works Branch, and as 'no attempt has been made by the architects at elaborate exterior ornamentation', the building was considered functional (WA Railway Gazette 1914, p. 7). The ground floor consisted of a library and reading room, a billiard room, a classroom, a secretary's office and a committee room. The first floor accommodation for the Technical School consisted of three large classrooms, separated by folding doors, so that when opened, the rooms converted into a large lecture hall. Detached from the main building there was also a timber and iron gymnasium (Swan Express 1914, p. 3).

At the time of its construction Midland was the only center in Western Australia, aside from Perth, Fremantle, Kalgoorlie and Boulder, to have such a facility. In 1904 there were only two technical classes with 16 students being held at Midland. By 1914, when the Railway Institute opened, there were 12 classes with approximately 150 students. In addition, 110–120 apprentices of the Midland Railway Workshops attended day classes where they were instructed in the intellectual side of their work. Classes included safe working, single and double-line operating, ticketing, vacuum brake instruction, telegraphy, shorthand and typing and first-aid training (McAllister & Heritage and Conservation Professionals 2009, p. 18). As the centre of technical training of Midland Railway Workshops apprentices there was always competition for places, apart from a decline during World War I when many men enlisted in the armed forces (McAllister & Heritage and Conservation Professionals 2009, p. 18).

An apprenticeship at the Workshops represented opportunity and security for the young working class in Midland. They entered at 15 or 16, dependent on their education, aptitude and contacts. Although there were mass retrenchments from 1930 to 1934, the darkest years of the Depression, employees of the Railways Department made good use of the educational classes run at the Institute (McAllister & Heritage and Conservation Professionals 2009, p. 19).

Few changes were made to the fabric of the Railway Institute during the early years, apart from minor external works in 1923 (McAllister & Heritage and Conservation Professionals 2009, pp. 19–20). In the 1970s and 1980s additional major renovations and repairs were carried out to the Railway Institute after the Technical School was relocated to the Midland Senior High School on Great Eastern Highway; after a period of use for off-duty staff, social and sporting events and library services, the building was closed in 1994 until its adaptation to offices for the Midland Railway Authority, which is discussed later in 'Current Uses and Sustainable Future Prospects'.

The Pattern Shop (1904)

The Pattern Shop was one of the original and core buildings on the Midland Railway Workshops site. The Pattern Shop was completed in 1903 under the supervision of The Chief Mechanical Engineer of the Midland Railway Workshops, E. S. Hume (Heritage and Conservation Professionals 2006, pp. 38–39). It is a substantial two-storey masonry building with a gabled roof sited to the west of Block Two, and was constructed in Federation Warehouse style with six structural bays and two pairs of doors along its northern elevation. It is a single volume internally with a timber-framed mezzanine, and three levels of timber-framed galleries in the Pattern Store at its western end (Heritage and Conservation Professionals 2006, p. 1). The pattern makers who worked in the shop were skilled timber and metal workers, crafting patterns that were precise models of various items to be cast in the foundry. Plans for patterns were carefully stored and their details indexed on cards stored in timber cabinets in the supervisor's office. The patterns themselves were stored on shelves in the Pattern Shop in case they were required in future (Heritage and Conservation Professionals 2006, p. 41). The Pattern Shop workers were a comparatively small proportion of the Workshops population, but their high level of skill afforded them both higher status and better wages than almost all other skilled tradesmen on the site (Bell 1990, pp. 29–30).

Minor changes to the Pattern Shop included the addition of a Foreman's Office, although its date of construction is unclear. Other changes were also additions, such as a large water storage tank to the original building, which remains unchanged internally, with patterns and machinery still housed within the Shop.

Similarly to the Railway Institute, it remained vacant for some time after the closure of the Workshops in 1994, until it was leased by the MRA as an atelier for designers and jewellers, although in the interim it was used as a storage facility for machines by the Machinery Preservation Club.

Closure of the Midland Railway Workshops

From the 1950s to the 1980s the needs of the workshops changed as a result of technological, social and economic changes, and as workers' skills became redundant, the functions of the various buildings were reassigned, some were demolished, and new ones erected. In 1984, the Commissioner for Railways, W. I. McCullough, announced the end of WAGR's association with light railways. From then on, demand for Midland Railway Workshops' facilities declined and, until its closure, the Workshops specialized in the manufacture of wagons for the bulk haulage trade and maintenance of freight wagons and the diesel–electric locomotive fleet of Westrail, the Western Australian Government railway system (Hocking Planning & Architecture 1995, p. 37).

In 1990 the Henshaw Report recommended rationalization of the Midland Railway Workshops; however, the government decided to close the Workshops, ending a 90-year role that they had played in railways operations (Heritage and Conservation Professionals 2006, p. 54). By early 1994 there remained fewer than 200 workers on site, and on 4 March 1994 the flag was ceremonially lowered, announcing the final closure of the Workshops (Ellis and Smyth 2004, p. 88).

The closure of the site was a devastating blow to the workers at Midland Railway Workshops since many of them had lost their jobs after close and long-term associations with the workplace. In the period following closure little was done at the site and the place was subject to neglect and decay (Hocking Planning & Architecture 1995, p. 38). In 2000 the Midland Redevelopment Authority was formed to manage the site and its future. After its establishment the MRA developed a plan for redevelopment of the site into a precinct comprising a mixed use of residential, commercial and light industrial use (Ellis and Smyth 2004, p. 100). In spite of virtually no start-up funding, the work to date was noted as an important case study by World Bank representatives at the International Conference on Impact Assessment held in Perth in 2008 (Moredoundt 2009, p. 4).

Current Uses and Sustainable Future Prospects

While the remaining significant buildings are being retained with the majority of their exteriors returned to their original state, the context of the site is currently undergoing change, through both the addition of new buildings and alterations to the landscape. The spaces between the buildings were bituminized, hardstand areas that were unforgiving, the pragmatic reality of an industrial site that functioned around the movement of large pieces of railway equipment and associated machinery. Remediation works to pedestrian paving between the buildings and areas of planted landscaping were essential to remove a meter of toxic substrate, which is in the process of being replaced with clean soil. The altered aesthetic is a clear compromise between retention of the original and the accessibility of the place to the general public for work, accommodation, recreation and education.

Blocks One, Two and Three

Blocks One, Two and Three are all currently unoccupied and are being maintained by the MRA to a minimum standard until appropriate future uses can be found. Given their sheer scale, a number of possible new functions have been proposed that leave the existing fabric largely untouched, with new works inserted into the interior. One potential function for at least one of the Blocks is as a center for tertiary education to service students along the eastern corridor of Perth, a demographic that is currently under-catered for. A general practice super clinic has been designed by Palassis Architects for insertion into Block One; it is essentially a highly contemporary two-storey 'building within a building', set back from the external skin to create an internal street, retaining the patina of the existing interior and allowing the original form, scale and fabric to be interpreted. An extrusion through the eastern end of the building is the only external indication of the clinic's existence. The architects also took advantage of the fact that, given that the roof of the original building would remain unaltered, the normal constraints of solar protection and drainage were eliminated, allowing for innovative design solutions that would not otherwise be possible (Moredoundt 2009). The outcome involves minimal change to the existing fabric, allows for interpretation, and is readily identifiable as new work.

Amy McDonnell, a Curtin University Interior Architecture student, proposed a venue for dance education, creation and performance in Block Two as part of her 4th-year dissertation in 2007. Her design approach:

Fig. 9.4 Interior Block Two of proposed studio insertion. (Designed and drawn by Amy McDonnell)

> … challenges the notion that heritage buildings impose severe restrictions that often result in a compromised design intent; instead this project explores the possibilities of creating a contrasting insertion within a building that successfully facilitates a new use while retaining the significance of the original structure. (McDonnell 2007, p. 3)

Philosophically, her proposal closely paralleled that of the super clinic design, an independent intervention that was clearly contemporary, and had minimal physical impact on existing significant fabric and incorporated interpretation, although it was on a much more modest scale, occupying less than one quarter of the floor area of the interior. McDonnell's selection of Block Two was driven by the freedom of the 'vast spatial volumes that are relatively obstruction free' and the potential of the contrast between raw industrial fabric juxtaposed with a sleek insertion that reflected the freedom and lightness of dance (McDonnell 2007, p. 21).

Through analysis of dance movement, precedent studies and the structure and spatial volumes of the building, her solution included a flexible dance space that could be reconfigured around varying performance types and audience requirements, a series of rehearsal spaces created from sliding panels suspended from the existing roof trusses, and mobile 'foldable' change rooms. Her response to the original building was to leave it relatively untouched:

> … the grimy, flaking paint on the walls; the patchy concrete floor; rust-marked steel columns; original machinery; tracks embedded in the concrete; lines painted on the floor; and pits dug out of the slab are all integral to the history and character of the building and have therefore been retained and worked around. (McDonnell 2007, p. 48)

She also created an extruded entry but through the existing doorway in the western end of Block Two, as an elevated lightweight structure suspended from the ground, in contrast to the robust original building. Although social sustainability is not overtly expressed as an agenda in the project, it is implicit in that it proposes a publically available venue close to the centre of Midland and accessible to the local population, many of whom retain strong associations with the place and their memories of its original use that, in her intervention, are retained in the fabric and form of the place (Fig. 9.4).

The Railway Institute

Refurbishment of the Railway Institute was completed in 2002, converting the building into administration offices for the MRA. The conservation plan policies for the place required minimal alteration both externally and internally and allowed no extensions. The philosophical approach taken by the architects, Philip McAllister Architect and Parry and White Architects in association, was to differentiate new works through use of contemporary style and materials based on Burra Charter principles (McAllister & Heritage and Conservation Professionals 2009, pp. 25–26).

Alterations to the building were limited, largely because the layout of the building comprized large rooms with an open central corridor and a centrally located staircase and were already well suited for their future use. Changes included lightweight glass partitioning and steel-framed stud walls to include amenities such as female toilets and tea preparation areas. Glass doors were inserted in the entrance as a security measure, but allowed the volume of the interior to remain evident. Unfortunately the building was broken into and set alight 3 days before construction commenced, all but destroying the central timber staircase. It was reconstructed to the original drawings and by comparison with an extensive photographic collection (McAllister & Heritage and Conservation Professionals 2009, p. 26).

Interpretive gestures were made in the form of stripping back a section of wall from floor to ceiling to reveal the original surface treatment and protecting it behind a perspex screen. A timber door was also stripped back to reveal the original fabric. It is interesting to note that the Railway Institute is the only project on the site so far to have a specialist practitioner in interior architecture, Lendis Golic, who was involved in the renovation of the interior, and was responsible for the finishes and fittings, including custom designed and manufactured carpet (Kinsella 2010, personal communication).

In terms of social sustainability, the conversion of the Railway Institute into an administrative center appears to be an appropriate solution, given how little change was needed to the existing significant fabric to accommodate a change in use, retaining much of the embodied memory that gives the place ongoing value. Its current function retains a link to its former purpose, in that it was originally a place that was dedicated to education, functionally separate from the other labor-oriented buildings on the site (Fig. 9.5).

The Pattern Shop

The Pattern Shop is the least altered of the buildings that have been adapted at the Midland Railway Workshops. The interior of the building has been retained intact, including machinery and the patterns in storage, and work was primarily confined to removal of material that was not significant. During the clean-up process a wall clad in corrugated iron was revealed and Palassis Architects, who designed and

Fig. 9.5 Interior Railway Institute. (Photographer Priya Metcalfe)

oversaw the works, argued for its retention even though its removal would have provided additional open work space. The concession was the insertion of a door opening with the remainder of the wall untouched, and unrestored, retaining its patina.

An extension to the eastern end of the Pattern Shop building in 2008 by Palassis Architects was a low-key, single-storey addition that allowed for retention of most of the significant fabric, and was driven by a philosophy of 'not making the building work too hard' (Palassis 2010, personal communication). Its scale is subservient to the original building while reflecting the overall form through proportion and roof pitch. It is steel-framed and metal-clad, in contrast to the masonry of the Pattern Shop, reflecting the practice of lightweight additions to core buildings that can be clearly distinguished from the original.

Its current use as an atelier of people that includes fine woodwork represents a continuation in its use as a craftsmen's workshop, since the original workers in the building were generally considered highly skilled and specialized artisans among the Railway Workshop population. The current occupants also focus on finely-crafted products including timber pieces and jewellery (Kinsella 2010, personal communication).

In 2005 Richelle Doney graduated from the Interior Architecture program at Curtin University, after completing a dissertation entitled 'Aged materiality—a world gone vintage' that proposed the adaptive re-use of the Pattern Shop into a wine emporium with both retail and museum functions, as a place of collaboration for smaller wineries in the nearby Swan Valley. The dissertation, while apparently primarily concerned with interpreting materiality both old and new, has at its core notions of social sustainability through community access and specialized use. Her proposed adaptive re-use:

> … revises the function of the building, while preserving the integrity of the architectural space. It ensures that our past is preserved for future generations, adding richness to our urban fabric; something that the community will identify with as well as recognise as part of our cultural history. (Doney 2005, p. 8)

Doney quite rightly recognizes that while the Pattern Shop is, by definition, a 'specific function holder', these kinds of industrial buildings have flexible spatial

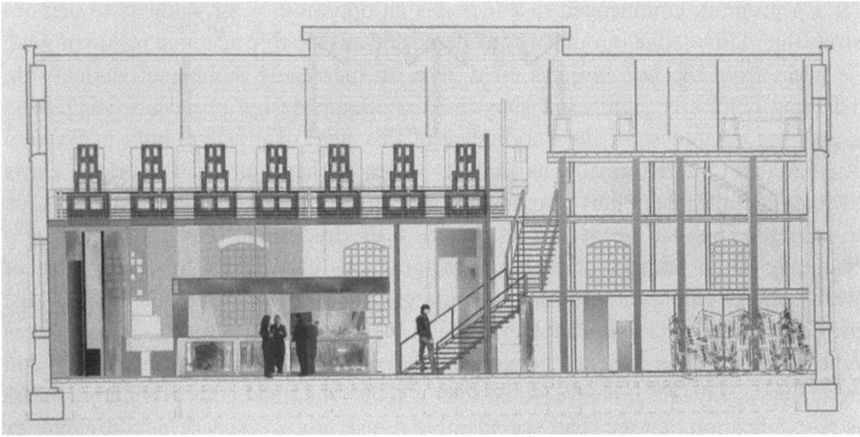

Fig. 9.6 Interior Pattern Shop proposal. (Designed and drawn by Richelle Doney)

qualities since they offer open internal spaces and generous heights (Latham 2000, p. 41). Her solution was developed through an understanding of the place, its history and its values to past users. The proposal was developed through a close investigation of the act of pattern-making, resulting in what was primarily a series of interchangeable semi-translucent screens within an open tasting bar and function area, and walk-through wine storage developed from an interpretation of the pattern racks. The interpretive gestures made allowed for a shift in function that revitalized the internal spaces while allowing the story of the place to be read in a contemporary reworking of the place's original use (Fig. 9.6).

Conclusion

In this chapter, two definitions of social sustainability are explored. One is of the social values that are acknowledged in the retention and ongoing use of culturally significant places and their settings as part of the continuum of human occupation. The other is the recognition of building-embodied histories and memories that communities value for their past associations and wish to preserve through conservation and, where necessary, adaptation, while recognizing that in such cases there is a degree of compromise to the use, fabric and settings of these places.

The dissertation proposals illustrate the potential for interior architecture to contribute to this important area of social sustainability. At Curtin University a core unit of study devoted to conservation and heritage practice in Western Australia has been embedded in the Interior Architecture 3rd-year program for the past 10 years. Students are exposed to the theories and practices of conservation, including ideas of memory and identity, history, culture, the Burra Charter processes, interpretation, archaeology, and conservation planning and works. Measured drawings and heritage assessments are used to examine various sites and their significance. In

2011 a program commenced that includes an opportunity for students to demonstrate their knowledge through a considered and sustainable adaptive re-use project.

Apart from the two projects cited here, an increasing number of students are selecting culturally significant sites for dissertation design proposals, suggesting a growing awareness of the importance of this field of practice, both in Western Australia and internationally, as students return to their countries of origin. However, an identifiable cohort of graduates specialising in the area in practice has yet to emerge. This can be in part attributed to the traditional stronghold held by a relatively small number of architectural practices in Western Australia, most of whom do not employ interior architecture graduates for conservation and heritage projects. As educators we shoulder some of the responsibility for this, since it is as undergraduates that students recognize areas of specialty such as conservation and heritage. As a result, it has yet to be seen whether the trend among undergraduates is any indication that we shall see adaptive re-use projects spearheaded by interior architecture practitioners with the specialist knowledge required to deal sensitively with important sites such as the Workshops.

As a site of exceptional cultural heritage significance with a diverse range of buildings and functions that have been superseded by changing technologies, the Midland Railway Workshops is no longer sustainable in its original use. The solution is its gradual conversion with the express intention of retaining its heritage values. The overall industrial nature of the site, particularly through the changes to the open spaces, will inevitably be altered in its character. While it could be argued that this conversion, along with the introduction of new buildings, will have a fundamental impact on the aesthetics of the Workshops, changing them from a gritty industrial workplace to a publicly accessible collection of open spaces that allow access to the site and its buildings, it represents the kind of compromise that has to be made to ensure the preservation and ongoing use of the place—and thus the retention of its social sustainability—albeit in an altered state.

References

Batterham, L. (2006). Women munitions workers, 1943–1945. In P. Bertola & B. Oliver (Eds.), *The workshops: A history of the Midland Railway Workshops* (pp. 128–141). Perth: University of Western Australia Press.

Bell, K. (1990). The Midland Railway Workshops 1920–1939. In J. Gregory (Ed.), *Western Australia between the wars 1919–1939. Studies in Western Australian history* (Vol. XI, pp. 29–30). Perth: University of Western Australia Press.

Doney, R. (2005). *Aged materiality—A world gone vintage*. (Unpublished Bachelor of Arts Interior Architecture dissertation). Perth: Curtin University.

Ellis, N., & Smyth, C. (2004). *Midland Railway Workshops: A history in pictures*. Western Australia: St George Books.

Heritage and Conservation Professionals. (2006). *Conservation plan pattern shop and store*. (Report For the Midland Redevelopment Authority).

Heritage Council of Western Australia (HCWA). (1992a). Register documentation, place number 03273 (pp. 2–3). http://inherit.stateheritage.wa.gov.au/Public/Search/PlaceNoSearch?placeNo=03273. Accessed 28 Aug 1992.

Heritage Council of Western Australia (HCWA). (1992b). Assessment documentation. Midland Railway Workshops. http://inherit.stateheritage.wa.gov.au/Public/Content/PdfLoader.aspx?id=b62651d2-6478-4e0d-a295-2a86d2f354d1&type=assessment. Accessed 28 Aug 1992.

Hocking Planning & Architecture. (1995). *Midland regional centre and environs conservation plan*. (Report for the Shire of Swan Australia Heritage Commission and the Heritage Council of Western Australia).

Hocking Planning & Architecture. (2009). *Block One conservation plan*. (Report for the Midland Redevelopment Authority).

International Council of Monuments and Sites Inc. (ICOMOS). (2000). *The Burra Charter*. Australia: ICOMOS. http://australia.icomos.org/wp-content/uploads/BURRA_CHARTER.pdf.

Latham, D. (2000). *Creative use of buildings* (Vol. 1). Dorset: Donhead Publishing.

McAllister, P., & Heritage and Conservation Professionals. (2009). *Conservation Plan Railway Institute*. For the Midland Redevelopment Authority.

McDonnell, A. (2007). *Space to move: Translating movement into an interior space to inspire dance creation and performance*. (Unpublished Bachelor of Arts Interior Architecture dissertation). Perth: Curtin University.

Midland Redevelopment Authority. (2010). *Midland Metro concept plan*. For Midland Central Redevelopment Area.

Moredoundt, N. (2009). *The challenges, conflicts and outcomes of managing twentieth century obsolescence in Western Australia*. Paper presented at ICOMOS 2009 conference. http://www.aicomos.com/2009-conference/. Accessed Jan 2011.

Murdoch University. (n.d.). *Midland Railway Workshops: An interactive oral history*. http://wwwmcc.murdoch.edu.au/midland/.

Swan Express. (1914, March 27). p. 3.

WA Railway Gazette. (1914, March 23). p. 7.

Chapter 10
Ambiguity and Permeability in Historic Modern Interiors: A Challenge for Cultural Heritage and Social Sustainability

Penny Sparke

Abstract The restoration and conservation of the interior spaces in nineteenth- and twentieth-century buildings (described collectively in this essay as 'modern interiors') play an important part in the contemporary world of cultural heritage and social sustainability. If these buildings are to take on lasting new roles that will make significant contributions to the lives of people who will visit or inhabit them, architects and interior designers, working with historians and other cultural heritage professionals, need to address the numerous subtle ways in which interiors have acquired, and continue to acquire, their multi-layered and constantly evolving definitions and meanings.

As they are also much more ephemeral than buildings, the restoration and conservation of historic interiors is often a more complex task requiring imagination and precise historical knowledge on the part of architects and designers.

This essay will address this complexity and its implications for today's architects and interior designers working in the areas of restoration and conservation within the context of cultural heritage. It will suggest that they need to be acutely aware of the past actual and potential psychological and socio-cultural meanings of interior spaces and to understand the ways in which they are both communicated by designers and read by users.

Keywords Domestic interior · Victorian home · Modern movement · Cultural heritage · Femininity · Machine esthetic

Introduction

The restoration and conservation of the interior spaces in nineteenth- and twentieth-century buildings (described collectively in this essay as 'modern interiors')—whether as museums or for re-use, either with the same or with a new purpose—play an important part in the contemporary world of cultural heritage and

---

P. Sparke (✉)
Kingston University, London, UK
e-mail: P.Sparke@kingston.ac.uk

D. Smith et al. (eds.), *Perspectives on Social Sustainability and Interior Architecture*,
DOI 10.1007/978-981-4585-39-2_10,

social sustainability. If these buildings are to take on lasting new roles that will make significant contributions to the lives of people who will visit or inhabit them, architects and interior designers, working with historians and other cultural heritage professionals, need to address the numerous subtle ways in which interiors have acquired, and continue to acquire, their multi-layered and constantly evolving definitions and meanings.

To date, the debates and discourses around the renovation and sustainability of buildings from the Modern era (1870–1970) have tended to focus on their role within urban renewal and regeneration and, where the interior is concerned, on notions of authenticity and appropriateness. This is especially so in the context of those countless spaces for which detailed information about their fittings, decor and furnishings has been lost. In their restored form they can only be approximations for what has been lost, or new interpretations relevant to the present. Indeed, as they are also much more ephemeral than buildings, the restoration and conservation of historic interiors is often a more complex task requiring imagination and precise historical knowledge on the part of architects and designers. Adding to this level of complexity, it could be argued that, inasmuch as they are not only defined technologically, visually, materially and spatially, but also psychologically and socio-culturally, the meanings of interiors, past and present, can be seen as being much richer than those of the buildings within which they were, and are, situated.

This essay will address this complexity and its implications for today's architects and interior designers working in the areas of restoration and conservation within the context of cultural heritage. It will suggest that they need to be acutely aware of the past actual and potential psychological and socio-cultural meanings of interior spaces and to understand the ways in which they are both communicated by designers and read by users.

This requires an understanding of the interior that goes beyond the need for authenticity or otherwise in the choice of interior components, materials and structural techniques and the ways in which (in, for example, a house museum) interior fittings were arranged and displayed. Further, in the case of a repurposed and restored modern interior, appropriate interior schemes and settings should be developed with an understanding, for example, of the ways in which gendered spaces have been created, read and understood, and how different identities have been, and continue to be, formed and confirmed in and by interior spaces. This kind of nuanced understanding of the ways in which interiors have worked in the past, and continue to work for their inhabitants and visitors is, I would like to suggest, often left to the historian to discuss in isolation. Most importantly, it requires the skills to understand the way in which the visual, material and spatial both form, and are formed by, the socio-cultural and the psychological.

Domesticity/Non-domesticity and the Interior

One of the key ways in which people and interiors have related, and arguably, continue to relate to interiors in the modern, industrial era was, and is, through people's perceptions of the level of domesticity present in an interior. The concept of domesticity has long since been linked to notions of gender (that of femininity in particular), class and individual/familial identity. It was embraced by the Victorians but repudiated by the Modernists, and it lies, it could be argued, at the heart of people's psychological relationship with interiors. Indeed, it was through the diametrically opposed readings of interiors as, on the one hand, 'comfortable and private' or, on the other, 'utilitarian and public'—or most commonly situated somewhere between those extreme ends of a spectrum—that most people negotiated, and continue to negotiate, their relationships with interiors. For the most part, however, the boundary between these oppositions was, and remains, blurred (Sparke 2008). Most modern and contemporary interiors displayed, and display, characteristics of both to a greater or lesser degree. This occurs whether they are the domestic spaces of house museums or restored dwellings, or the predominantly non-domestic interiors of nineteenth- or twentieth-century warehouses, retail outlets, buildings linked to transportation and leisure, institutions, hospitals and prisons. In making historical modern interiors socially sustainable, I would suggest, architects and designers need to understand the way those levels of ambiguity and permeability operated in the past, and address the subtle ways in which renovated modern interiors can express differentiated levels of both domesticity and non-domesticity within them, as well as embodying the inherent tensions created by their coexistence.

Ambiguity, Permeability and the Modern Interior

This essay will focus on some specific questions that can arise in any restoration-conservation project; namely, where does the visual, material, and spatial language of domesticity start and end in any of the interiors described above? That is, how much domesticity is present in a nineteenth-century home as compared to a prison or lunatic asylum (where one would expect very little but where some vestigial signs may remain); how is that domesticity communicated and recognized; and how can, or should, it be represented visually, materially and spatially? And, secondly, to what extent does the exterior world enter into the interior or, alternatively, do interior spaces leak into the world outside the buildings that house them, thereby blurring the edges of the relationship between the worlds of privacy and publicity, of domesticity and non-domesticity? Inasmuch as, frequently, the extension of markers of domesticity beyond the interior proper, or the penetration of the public sphere into the home, introduces ambiguities, these questions are closely linked. In the context of restoration and conservation, these ambivalences and ambiguities make it difficult to define both the nature and the boundaries of the interiors that are being addressed and interrogated, thereby adding a level of complexity to the task at hand.

These questions preoccupy the historian of the modern interior. However, given that the way in which these characteristics of the interior have been conventionally expressed, they continue to be relevant into the present, and visitors to historical modern buildings and their interior spaces, and the occupants of contemporary repurposed interiors, respond instinctively to the signs of domesticity and non-domesticity that exist within them, as well as to the ways in which the inside relates to the outside. The questions are also relevant to debates within contemporary cultural heritage and social sustainability. An awareness of the way in which these characteristics have been reconstituted in a heritage site undoubtedly enhances the experience of the visitor within it, taking it beyond a mere response to the aesthetic of the interior in question, towards a deeper understanding of the psychological and socio-cultural issues emanating from it. The transformation of a former industrial warehouse into apartments, for example, will need to take into account the ways in which future inhabitants will understand how their public and private selves can be accommodated within their ambiguous and radically transformed spaces. With that in mind, this essay will focus on the questions posed above and discuss some of the issues raised by them, with reference to specific examples.

The 'Degree Zero' of Domesticity

I would like to start with a discussion of what we might call, using a term coined by the French cultural critic, Roland Barthes, the 'degree zero' of the domestic interior (Barthes 2010). A sight that can frequently be found inside North American, and increasingly, in European shopping malls is one of little oases of domesticity, intended for rest and respite. They consist of a few inwardly-oriented chairs, arranged for conversation, positioned on a rug and framed by potted plants—whether real or artificial—which separate them from the larger interior space that embraces them. These spaces within spaces are the 'degree zeros' of domesticity. They contain just enough elements to communicate a message about comfort, rest, privacy and a moment's freedom from the demands of the commercial, industrialized world and those of the public sphere. In material and spatial terms their key elements are comfortable, upholstered seating; an oriental rug which, softening the industrial material of the mall floor around it, evokes the craft skills of the past and the exoticism of faraway lands; and potted plants, which act as boundary-defining screens but which also represent the presence of tamed nature in these otherwise heavily acculturated spaces. Their exotic tropical origins confirm the power of escapism offered by these little settings, which assert themselves in opposition to the interior values of the mall itself. The widespread presence of contemporary oases such as these, in addition to those frequently found in spaces such as bookshops and hospital waiting rooms, suggests that the language of domesticity is well established and easily learnt, understood and reproduced; that its presence beyond the walls of the home is deeply embedded within modern industrial culture; and that it is easily recognizable even in its 'degree zero' form. While on one level this is very obvious, on another it demonstrates the subliminal way in which interior spaces communicate with their inhabitants, and the importance of our shared understanding of what constitutes the presence of domesticity in a space.

Victorian Domesticity

The contemporary oasis described above is totally dependent on a visual, material and spatial language that was constructed at least a century and a half ago. The private face of the modern interior was formed in the middle-class Victorian home. That space performed a number of simultaneous roles, primary among them being that of a comforting refuge from the worlds of work and commerce. As the nineteenth century progressed, however, the Victorian home also became a focus for aesthetic intervention, a destination for goods acquired through mass consumption, and a focus for media attention. As well as offering opportunities for private social activities and interactions, the nineteenth-century domestic arena also provided possibilities for social display. Above all, the ideology of the separate spheres made it a site where modern feminine subjectivity was largely negotiated, where fashionableness could be demonstrated, and where women could express themselves through their new role as 'decorators' (Sparke 2010).

Typically, the spaces of Victorian domesticity, which provided a model for subsequent developments, were filled to the brim with items of comfortable, upholstered furniture, textiles on every available surface, bibelots on the mantelpiece, patterned carpets, and potted plants. In a fairly typical middle-class living space in Manchester, England, from the last years of the nineteenth century, a crowded mantelpiece, multiple framed pictures suspended from the picture rail, prominently displayed plants in pots, an eclectic mix of chairs arranged semi-formally around little tables and a large, religious statue positioned in one corner combined to create an impression of an inward-looking home dedicated to comfort, self-reflection, social interaction and private religiosity. By the second half of the nineteenth century, that language of Victorian domesticity had become a feature of the interior landscapes of a range of semi-public and public 'homes from home', including cafés, restaurants, women's clubs, hotels, the leisure areas of department stores, railway waiting rooms and carriages, and mental hospitals. That replication challenged the idea of the separation of the spheres, such that modernity, and by extension the modern interior, were ultimately defined by the crossover between the private and the public arenas rather than the distinction between them (Sparke 2010).

By the early twentieth century, that crossover had been reversed by the reintroduction of public arena spaces into the dwelling. For many of the Modern Movement architects of that era, the interior became so inextricably linked with Victorian middle-class domesticity, the psychological make-up of its occupants, and the materialistic world of mass consumption, that they felt compelled to develop an architecture that minimized its existence. They found an alternative model, which they believed to be both rational and functional, in the spaces inside the new public sphere buildings—factories, stores and exhibition halls among them. They were also inspired by the functional spaces in new objects of transport, including Pullman train kitchens and ships' galleys (Collins 1965). Several focused on social housing projects and the idea of the 'minimal dwelling', but many of their commissions came from progressive, middle-class clients who wanted a taste of what was rapidly to become a new, clutter-free lifestyle.

In spite of the numerous hesitations expressed about it, a Modern Movement domestic interior inevitably emerged. Unlike its nineteenth-century predecessors, it was primarily spatially defined. Its roots lay in what the Modernists believed to be the unconscious, utilitarian, 'engineered' aesthetic of the new public sphere interiors, and they introduced it into modern residential spaces in their efforts to address what they saw as the 'problem' of bourgeois domesticity. Rapidly, however, it was recirculated back into the public arena and applied to a wide range of building types—restaurants, shops, leisure centres, schools, hospitals and churches among them—some of which were new and others of which were being 'modernized' for the first time. In that new context it became a much more self-conscious aesthetic which openly declared its alliance with modernity. In effect, once it had been reformulated within the Modernist dwelling, the modern public interior was transformed into a metaphor for itself. The Modern Movement architect-designers also embraced the idea of the *Gesamtkunstwerk* (the whole work of art). Architecture took the lead, while the so-called 'decorative arts', a catch-all term which embraced the interior, followed meekly after it. For the Modernists the interior was simply the space within buildings, an inevitability which, in order for daily life to take place, had to be 'equipped', albeit as minimally as possible.

The Modernists' rational approach to space planning inevitably impacted most strongly on those areas of the house dedicated to work, rather than to leisure, display, social relations or interiority. That was especially the case as household servants became increasingly scarce and the housewife had to take on more household tasks. In its early formulation, the domestic rational interior focused exclusively on process rather than aesthetics. In the hands of Modernist architects, however, a characteristic aesthetic for the interior also began to emerge. In line with modernity's prioritization of the visual, by the inter-war years the rational interior had become increasingly associated with a simple, abstract, geometric, undecorated interior style, often referred to as the 'machine aesthetic'. It denoted a high level of functionality and came to be seen by many as *the* modern interior aesthetic of the twentieth century. Although the style was initially only applied to domestic kitchens, bathrooms and other work-oriented areas, as it increasingly came to represent modernity itself, and, by extension, the modern lifestyle, it quickly spread to other areas of the house as well, the living room and dining room among them. The effect was yet another blurring of the distinction between the appearance of interiors outside and inside the home. The function-driven, rational, essentially non-domestic—or, as it became in the hands of the Modernists, 'anti-domestic'—approach to home interiors appealed to housewives seeking to put their domestic role on a professional footing, to a group of early twentieth-century feminists who sought to develop collective housing as a means of supporting a way of life which rejected the Victorian ideology of domesticity, and to politically-motivated Modernist architects and designers who set out to develop standardized social housing projects which would provide large numbers of people with access to basic living standards (Bullock 1988).

An example of a Modern Movement interior, destined for private dwellings, that overtly brought the rational, public world of the factory inside and created a level of ambiguity in the modern home that has never left it, was the inspiration of a German architect, Ernst May. He was involved in a huge redevelopment of Frankfurt. Impressed by what he saw at the *Weissenhof Siedlung* (a housing estate designed to showcase design) of 1929, he initiated a thorough program of research as part of the Frankfurt project (Bullock 1988). His design team studied psychology, material and product evaluations, and, of course, scientific management principles as applicable to the home. They scrutinized every aspect of household design, in order to produce efficient and content housewives: 'color brightened the housewife's world, making housework more tolerable; enamelled surfaces made for easy cleaning; and furniture with smooth lines eliminated dusting in hard-to-reach places', one writer has explained (Coleman et al. 1996, p. 232). It was beginning to become clear how a rational approach to home design could take on visual and material forms to represent it. One of May's most important decisions was to bring the architect Margarete Schütte-Lihotzky from Vienna to work with him on his project. Her lasting contribution to the development of the rational face of the modern domestic interior was the work she undertook on what came to be known as the 'Frankfurt Kitchen'. The design of the small laboratory kitchen that Schütte-Lihotzky developed, along lines already set out by the American protagonist of scientific management in the home, Christine Frederick, was inspired by equivalents in the public sphere. These were the ship's galley, the kitchen in the railroad dining car, and the lunch wagon in particular, which had been designed to facilitate serving food to large numbers of people in as efficient a way as possible (Coleman et al. 1996, p. 235). The architect emphasized the importance of step-saving and of efficient, well organized storage. She also included Frederick's workbench and stool in what has been described as 'a workstation where all implements were a simple extension of the operator's hand' (Coleman et al. 1996, p. 235). Other notable features included a continuous counter surface attached to the walls of the tiny room, a cutting board fitted into the workbench with a waste bin positioned immediately below it, and a wooden plate holder fixed to the base of the glazed wall cupboards. Schütte-Lihotzky's small 'laboratory' kitchen transformed the American Christine Frederick's earlier, essentially utilitarian, spaces into a much more visually elegant solution to the same problem (Frederick 1913) (Fig. 10.1).

While so much of Schütte-Lihotzky's design depended on the ideas and actions of Christine Frederick before her, it had a more 'modern' look and feel to it than the more overtly functional kitchens the latter had created (Beecher and Stowe 1869). That was achieved through the visual integration of the elements into a whole, a self-conscious use of colour. The cabinet fronts were a deep blue (although even this could be justified functionally as it was a colour which was believed to repel flies), and strikingly modern materials, including linoleum, glass and metal, were employed. While they could also be justified functionally, they made a significant contribution, nonetheless, to the kitchen's modern look. The impact of the Frankfurt Kitchen, which quickly went into mass production, was enormous. It took the idea of rationality into the home on a significant scale and, through the control it gave to her, became a symbol of the professionalization of the housewife. At the same time it removed much drudgery from food preparation. The same principles were soon

Fig. 10.1 The 'Frankfurt Kitchen', designed by Margarete Schütte-Lihotzky 1926–1927. (Image courtesy of the University of the Applied Arts, Vienna)

applied to the bathroom and to the laundry. The emphasis on the kitchen, however, had implications for the conceptualization of the modern house or apartment as a whole, such that, in the words of Schütte-Lihotzky herself, 'the arrangement of the kitchen and its relationship to the other rooms in the dwelling must be considered first' (Beecher and Stowe 1869, p. 187). That seemingly simple statement had dramatic implications for the evolution of the modern domestic interior, as it suggested that houses should be designed from the inside out, rather than the reverse. The strategy was soon to become one of the lasting tenets and legacies of architectural Modernism. While, on the one hand it had the effect of minimizing the autonomy of the interior by merging it with the architectural shell, on the other, the idea of letting the plan determine the façade imbued the interior with a new level of importance. By giving the elements within the plan utilitarian definitions—the kitchen, the bathroom and the living space for example—they became driving forces behind architectural design as a whole. Indeed, the interior became the pivot around which all architectural decisions were made. A restored original version of the Frankfurt Kitchen was displayed at the Victoria & Albert Museum *Modernism* exhibition, held in London in 2005.

Conclusion

In the context of the restoration and conservation of historic Modern Movement interiors as heritage sites and as repurposed buildings, an awareness of the degree to which they can be seen to be communicating ideas about domesticity or non-domesticity, the private or the public, is crucial. It also demonstrates to visitors, users and inhabitants the fact that we continue to blur the edges of our private and public worlds and that there is something deep within us that wants, simultaneously, to define and express our own individuality while also subjugating the spaces we inhabit to the rational work ethic that we still see as part of progress and control. It is a tension that lies deep within us and is represented, as it was in the nineteenth and twentieth centuries, within our contemporary interior environments.

Whereas, as we have seen, the nineteenth-century domestic interior was largely inward-looking, the Modern Movement architects reversed that, creating, in an attempt to deny the existence of domesticity, residential interiors which were outwardly-oriented. The restoration of modern interiors of all kinds needs to be sensitive to these intentions and to the numerous strategies that were used to achieve the required ends. From an architectural perspective they include the construction of what may be called transitional spaces. In the nineteenth century, conservatories attached to parlours, balconies and bay windows served to bring the outside in through the inclusion within them of plants and flowers, and sometimes birds and fish as well. This taming, or domestication, of nature within these transitional spaces of the nineteenth-century home was reversed by the Modernists. Their inclusion of nature within inside spaces served, rather, to take the eye out of the (usually) open plan interior towards nature in its raw state outside. To this end, balconies and terraces were widely used and, occasionally, winter gardens were also introduced. For a description of Ludwig Mies van der Rohe's Villa Tugendhat in Brno, for example, see Hammer-Tugendhat and Tegethoff (2000, pp. 2–3).

This journey back into the history of the modern interior makes several things clear. Firstly, the relationship between domestic and non-domestic, private and public spaces—and the all-important shades of grey in between—as well as the blurring of the boundary of what constitutes inside and outside space, is highly complex and continually evolving. Inevitably that tension continues into the present. The private home that has become a house museum, the warehouse and chapel that have been transformed into private dwellings, make the issue even more complex today. Historians have worked to understand the reasons for this working with the visual, material and spatial evidence in front of them, also drawing extensively on contextual historical material and relevant contemporary theory. Increasingly, architects and interior designers are working with historians and other cultural heritage practitioners to search for new interpretations and approaches. The continued collaboration between them should prove extremely rewarding and offer new ways of bringing historical work in this area to the debates around cultural heritage and social sustainability.

References

Barthes, R. (2010). *Writing degree zero*. London: Modern Classics.

Beecher, C., & Stowe, H. B. (1869). *The American woman's home*. New York: J. B. Ford and Co.

Bullock, N. (1988). First the kitchen—then the façade. *Journal of Design History, 1*(3 & 4), 37–56.

Coleman, D., Danze, E., & Henderson, C. (Eds.). (1996). *Architecture and feminism*. New York: Princeton Architectural.

Collins, P. (1965). *Changing ideals in Modern architecture 1750–1950*. London: Faber and Faber.

Frederick, C. (1913). *The new housekeeping: Efficiency studies in home management*. New York: Doubleday.

Hammer-Tugendhat, D., & Tegethoff, W. (2000). *Ludwig Mies van der Rohe: The Tugendhat House*. Vienna: Springer.

Sparke, P. (2008). *The modern interior*. London: Reaktion Books.

Sparke, P. (2010). *As long it's pink: The sexual politics of taste*. Nova Scotia: University of Nova Scotia Press.

Index

D. Smith et al. (eds.), *Perspectives on Social Sustainability and Interior Architecture*,
DOI 10.1007/978-981-4585-39-2,

MIX
Papier aus verantwortungsvollen Quellen
Paper from responsible sources
FSC® C105338

If you have any concerns about our products, you can contact us on
ProductSafety@springernature.com

In case Publisher is established outside the EU, the EU authorized representative is:
Springer Nature Customer Service Center GmbH
Europaplatz 3, 69115 Heidelberg, Germany

Printed by Libri Plureos GmbH
in Hamburg, Germany